Christiana Figueres was the UN Executive Secretary for Climate Change between 2010 and 2016. She was the public face of the most pivotal climate agreement in history, the Paris Climate Agreement, in 2015. Christiana is a council member of The Earthshot Prize, the most prestigious global environment prize in history.

Tom Rivett-Carnac was senior political strategist for the Paris Agreement. Together they are the co-founders of Global Optimism, an organisation focused on creating environmental and social change, and co-hosts of the podcast Outrage and Optimism.

THE FUTURE WE CHOOSE

Christiana Figueres
Tom Rivett-Carnac

The Stubborn Optimist's Guide to the Climate Crisis

MANILLA PRESS

First published in the UK by Manilla Press
An imprint of Bonnier Books UK
Wimpole Street, London, W1G 9RE
Owned by Bonnier Books
Sveavägen 56, Stockholm, Sweden

Hardback – 978-1-838-770-82-2
Trade paperback – 978-1-786-580-36-8
Paperback – 978-1-786-580-37-5
Ebook – 978-1-838-770-83-9

A CIP catalogue of this book is available from the British Library.

Typeset by Richard Marston
Printed and bound in Great Britain by Clays Ltd, Elcograf S.p.A.

1 3 5 7 9 10 8 6 4 2

Every reasonable effort has been made to trace copyright holders of
material reproduced in this book, but if any have been inadvertently
overlooked the publishers would be glad to hear from them.

Manilla Press is an imprint of Bonnier Books UK
www.bonnierbooks.co.uk

*We dedicate this book to Christiana's daughters,
Naima and Yihana, and Tom's daughter and son,
Zoë and Arthur, and to the generations who
will inhabit the future we choose.*

Contents

'Let us not pray to be sheltered from dangers, but to be fearless when facing them.'

Rabindranath Tagore

Authors' Note

We are good friends and fellow travellers on this planet, but we differ in many ways. We were born in two different geological periods. Christiana was born in 1956, at the end of the 12,000 year Holocene epoch, when a stable climate allowed humanity to flourish, and Tom in 1977, when the Anthropocene epoch – characterised by humanity's destruction of the very conditions that allowed us to thrive – began.

We come from opposite sides of the geopolitical map; Christiana from Costa Rica, a small developing country that has long been a model of economic growth in harmony with nature, and Tom from the UK, the world's fifth-largest economy and the birthplace of the Industrial Revolution and its reliance on coal.

Christiana comes from a deeply political family with immigrants to Costa Rica on both sides. Her father was three times president of the country and is considered the father of modern Costa Rica. Not only did he initiate some of the most far-reaching environmental policies in the world, he remains the only head of state ever to have abolished a national army. Tom stems from a family steeped in British history and rooted in the private sector. He is a direct descendant of the founding chairman of the East India Company when it was the only company in history to have a private army. Tom's earliest memories are of looking for oil with his petroleum geologist father.

Christiana is the mother of two adult daughters, and Tom is the father of a daughter and a son, both under age ten.

We could have had nothing in common, but we deeply share that which is most important: concern for the future of our children and *yours*. In 2013, we decided to work together to forge a better world for all children.

From 2010 to 2016, Christiana was Executive Secretary of the United Nations Framework Convention on Climate Change, the organisation tasked with guiding the response of all governments to climate change. Assuming the highest responsibility for negotiations right after the dramatic debacle of the 2009 Copenhagen climate change conference, Christiana refused to accept that a global agreement was impossible.

In 2013, she heard about Tom, who was then president and CEO of the Carbon Disclosure Project USA. and a former Buddhist monk. Intrigued by his unusual combination of experiences, Christiana asked him to join her in New York City to discuss the possibility of him joining the UN.

At the end of a walk around Manhattan that took the better part of the day, Christiana turned to Tom and said,

'It's clear to me that you have none of the experience necessary for this job. But you have something far more important: the humility to foster collective wisdom, and the courage to work within a complexity that is beyond any mapping.'

With that, she invited him to join the UN effort to advance the negotiations for the Paris Agreement as her Chief Political Strategist. He designed and led the largely covert Groundswell Initiative, which mobilised support for the ambition of the agreement from a wide range of stakeholders outside of national governments. A few years later the most far-reaching international agreement on climate change ever attempted was finally achieved.

When the green gavel came down at 7:25 p.m. on 12 December 2015, adopting the Paris Agreement, 5,000 delegates who had been holding their breath for hours jumped out of their seats in ecstatic delight, in celebration of the historical breakthrough. One hundred and ninety-five nations had just unanimously adopted an agreement to guide their economies for the next four decades. A new global pathway had been charted.

But pathways are valuable only if they are used. Humanity has procrastinated for far too long on climate change – now we have to walk the path, or rather we have to run it. This book maps the route of that run, and we hope you will run alongside us. Join us at www.GlobalOptimism.com.

The Critical Decade

We wrote this book before COVID-19 disrupted our world. In fact, we had only managed the first three stops on a planned year-long book tour before we rushed to our respective homes in Costa Rica and the United Kingdom, and into a global lockdown. Since then we have been shocked at how many aspects of both the dystopian and the desirable future we describe in this book suddenly came into relief and stark contrast with each other. More than ever we are determined to play our part in ensuring our future is one that we all deliberately choose, rather than one we stumble blindly into.

The beginning of this decade has thrust upon us levels of unparalleled intensity. Whether we're experiencing loneliness, fear, grief, excitement, hope or gratitude, we've had to adjust to a state of heightened sensibility, with two competing realities jostling for our attention.

One reality is the relentless depletion and degradation of our global commons – our forests, oceans, rivers, soil and air – even though we know our health and well-being rely upon them. We're seeing the continued pursuit of economic growth through the unbridled extraction and burning of fossil fuels, despite the fact that we know this is changing the chemistry of our atmosphere, heating our planet and pushing the earth systems that sustain us to breaking point. The decade began inauspiciously, with the deadly COVID-19 pandemic, lockdowns, school and workplace closures distracting us temporarily from the longer-term challenges. A stark reminder that those longer-term challenges remain in place is the fact that while 2020 saw a marked decrease in greenhouse gas emissions, at the same time it broke the record for the hottest year on the planet.

While many people remain unaware of the scale of the ongoing and intense destruction, and some even choose to ignore it, everyone is now beginning to feel the consequences. Species extinctions, super storms, heatwaves, droughts, fires and the human suffering and economic damage they cause are increasing in frequency, compounding centuries of inequality and human rights atrocities as the roots of political and social unrest. We may think of these issues in silo form, but they are all intricately interconnected.

We cannot close our ears or turn our eyes away from all the pain. We also cannot ignore the fact that, should we continue on the business-as-usual path, we may be orchestrating the demise of our own species. We still have not sufficiently connected the dots between the ongoing destruction of our natural habitats and our future ability to ensure our own and our children's health and safety, to feed ourselves, inhabit coastlines, and uphold the integrity of our homes.

It's a difficult reality to swallow, but we need to. If we don't, we will not be able to empathise with the despair that so many people cannot, understandably, move beyond.

In equal standing, we have to courageously hold our conviction that despite that reality, and in fact maybe even because of it, we do have the potential to activate ourselves in an opposite direction, and that this is already starting to happen. We are now seeing an intense series of responses to the climate and planetary crises beginning to unfold in communities, in companies, in cities and even in national governments, spurred on by the ever more alarming scientific data, and also by people from all walks of life calling out for the changes we so urgently need.

We remember a twelve-year-old girl marching with her friends down 16th Street in Washington D.C. at 10:00 a.m. on a Friday, holding up a hand-painted sign of the Earth enveloped in red flames. In London, adult demonstrators dressed in black and wearing riot-police headgear formed a human chain blocking traffic at Piccadilly Circus, as others glued themselves to the pavement in front of the headquarters of BP. In Seoul, the streets teemed with elementary schoolchildren sporting multicoloured backpacks and carrying banners that said CLIMATE STRIKE – in English, for the benefit of the media. In Bangkok, hundreds of teenage students took to the streets – with firm resolve and heavy hearts, they walked behind their defiant leader, an eleven-year-old girl carrying a sign: THE OCEANS ARE RISING AND SO ARE WE.

From the quest for independence in India to the civil rights movement in the United States, civil disobedience has often erupted when a reigning injustice becomes intolerable, as we are now seeing with climate change. The pain and torment of this moment, the unacceptable generational injustice and deplorable lack of solidarity with the vulnerable have already opened the floodgates of protest. This protest – from young people online or marching in the streets, from changing customer and shareholder demands, from lawsuits, boycotts, and also from voters at the ballot box – is propelling climate action and awareness to new levels. Combined with rapidly changing economics that make the solutions to the climate crisis ever more attractive, this rising tide of activism gives policy makers a strong mandate to reflect and further enact the political and systemic changes we need.

In addition, the positive effect of the historic Paris Agreement, which all governments of the world unanimously adopted in December 2015, and most ratified into law in record time, is undeniable. The Agreement delineates a unified strategy for combating climate change, and today every major power in the world is already planning to completely transition its energy system to renewables. Major economies, including China and the US (where President Biden has put climate at the forefront of his administration's agenda, testified to by his re-joining of the Paris Agreement on his first day in office), have committed to reaching net-zero emissions by around the middle of the century, as have well over a thousand big corporations. Some companies and governments are planning to get there well before 2050, and some already have. Oil and gas companies are being forced to reconsider their futures on time frames previously inconceivable, in part because of the pandemic-related demand slump – but also because the alternatives are rapidly becoming less risky and increasingly competitive. Coal has become an untenable investment for most major financiers, as solar and wind are now the cheapest sources of new electricity generation in most countries around the world. Money is moving decisively away from high-carbon investments to low-carbon investments. We are firmly on track, albeit at the beginning of that track, to completely transform the way we produce and consume energy, and that in turn is already propelling profound changes in our industrial, transportation and agricultural sectors too.

There are many people for whom these transformations are not coming fast enough, and for whom the incremental nature of goal- and target-setting seems woefully inadequate given the scale of our crisis. After all, we have known about the possibility of climate change since at least the 1930s and have been certain since 1960, when geochemist Charles Keeling measured CO_2 in Earth's atmosphere and detected an annual rise.[1] While most governments dithered, behind the scenes environmentalists and climate activists were working hard to lay the foundations for the necessary change, and at last the ground is rich and strong enough for a burst of exponential activity that can drive solutions at the pace we need. Change typically happens gradually, then suddenly, and the 'suddenly' part of climate action is at last beginning to blossom, evident in the early unfolding of the most exciting economic transformation we have seen in our lives.

These two competing realities, with their respective potential futures – one dystopian and one regenerative – have equal momentum now, even if most people currently think the first more likely. If we were to visualise these two realities as time lines on a graph, we believe that this moment, the beginning of this critical decade, is when we finally arrive at the cross-over point. This is when the rising momentum for protecting and restoring our global commons at last outpaces the reality characterised by their destruction. And it is the sheer intensity of these two possible trajectories that makes this a uniquely exciting and privileged moment to be alive – bewildering and exhilarating at the same time.

Our responsibility now is to fertilise the trajectory of the future we do want, and there's never been so much wind at our back. We have already achieved a host of social and political successes; we have most, if not all, of the technologies we will need; we have the necessary capital; and we know which policies are most effective. The changes we need to make are significant, but we can make them.

If we were to stand in the future and look back at this decade, as historians have for example viewed the Renaissance, or the Enlightenment or the Digital Revolution, we would see that that this is a true turning point: the point at which (building on the foundations of reason, science, technology and humanist philosophy) we have the opportunity to fully embrace our interdependence with all of nature and each other, and deliberately and intentionally change course.

This is the moment when greenhouse gas emissions from human activity begin their descent, and with that descent comes a growth of new jobs and improvements to our health: we'll have better energy and food security, cleaner air, thriving biodiversity and human prosperity. This is the moment in which we finally figure out that we actually do love life, ourselves and each other enough to save ourselves.

The two of us stand with one foot in outrage and the other in optimism, our heads buzzing with excitement for what can still be achieved. We invite you to stand with us – acknowledging the two realities in front of us – and to play your part in making possible the change we need by embracing stubborn optimism (and we've dedicated a whole chapter to outlining exactly how to be stubbornly optimistic).

How we achieve well-being for everyone everywhere on a thriving planet will be the most poignant chapter of the story of our species. This book is a guide to how we can write that exhilarating chapter together.

PART I
TWO WORLDS

CHAPTER ONE

Choosing Our Future

Geological time is long and slow. Or at least it used to be. Ice ages, during which vast glaciers covered much of the northern continents, have sluggishly come and gone throughout the history of our planet. The last ice age lasted about 2.6 million years. With very gradual warming resulting from natural influences on Earth's climate, we slowly left that ice age and entered the Holocene epoch, which stretched out over 12,000 years – until the twentieth century – under relatively stable temperatures, fluctuating only 1 degree Celsius above or below the average.[1]

Throughout that geological period, temperatures, precipitation patterns, and terrestrial and ocean ecosystems settled into a 'sweet spot' of natural conditions conducive to human propagation and well-being. That environmental stability allowed the human species of approximately 10,000 people living in small tribes to start a sedentary life, evolve into agricultural farmers and settlers, and eventually develop cities, supported by industry and machine manufacturing.

It allowed humans to thrive and the population to grow to the current 7.7 billion.[2]

During the Holocene, 'life created the conditions conducive to life.'[3] And we could have continued in that geological era. But we didn't.[4]

Over the past 50 years, we have severely undermined the environmental integrity of our Blue Marble and threatened our continued life here. Our post-Industrial Revolution lifestyles have caused massive damage to all our natural systems. Mainly because of the unbridled use of fossil fuels and vast deforestation, the concentration of greenhouse gases in the atmosphere today exceeds anything we have had since well before the last ice age,[5] resulting in extreme weather events of increasing frequency and intensity all over the world: floods, heatwaves, droughts, wildfires and hurricanes. Half the world's tropical forests have been cleared, and every year about 12 million more hectares are lost. In about forty years, at the current rate, 1 billion hectares could be gone – a land mass equivalent to Europe.[6] In the last 50 years, the populations of mammals, birds, fish, reptiles, and amphibians have, on average, declined by 60 per cent. Some suggest we are already living through the sixth mass extinction.[7] According to the latest research, 12 per cent of all surviving species are currently threatened, and climate breakdown will significantly amplify that threat.[8] Oceans have absorbed more than 90 per cent of the extra heat we have produced over the last 50 years.[9] As a result, half the world's coral reefs are already dead,[10] and the Arctic summer sea ice, whose reflective capacity helps to regulate temperatures all over the world, is shrinking rapidly.[11] The melt from land glaciers has already caused sea levels to rise more than 20 centimetres, leading to major salt intrusion in many aquifers, worsening storm surges and existential threats to low-lying islands.[12] In short, in just the last 50 years we have catapulted humanity and the planet out of the previous benevolent Holocene epoch and into the Anthropocene, a new geological period in which biogeochemical conditions are dominated not by natural processes but by the palpable impact of human activity. Humans are for the first time ever the prime driver of large-scale climate change on the planet.[13]

All studies you may read about the Anthropocene epoch point to the unprecedented levels of destruction that we have caused in just five decades.[14] The underlying assumption in those analyses is that we have irretrievably cast our die and that increasing destruction will be the leitmotif of the entire geological era.

We take a radically different view.

We argue that devastation is admittedly a growing possibility but not yet our inevitable fate. While the beginning of this period of human history has been indelibly and painfully marked, the full story has not been written. We still hold the pen. In fact, we hold it more firmly now than ever before. And we can choose to write a story of regeneration of both nature and the human spirit. But we have to choose.

In deciding what kind of world we and future generations will live in, we don't have many options; we have in fact only two, both of which are set out in the Paris Agreement, and both of which we present here for your consideration. Keep in mind that we have already warmed the planet by 0.9 degrees Celsius more than the average temperature before the Industrial Revolution. Under the Paris Agreement, all nations committed to collectively limit warming to 'well under 2 degrees Celsius', and ideally no more than 1.5 degrees Celsius (2.7 degrees Fahrenheit), through national emissions-reduction efforts that substantially increase every five years. To start the process, in 2015, 184 countries registered details of what they would do in the first five years and agreed to come back every five years to make stronger commitments, since the first round of commitments was only the first step toward achieving the long-term goal of net-zero emissions.

We present two scenarios. One or the other will become our reality.

The world we are now creating, leading to warming of more than 3 degrees Celsius.[15] The first scenario we set out illustrates the very dangerous trajectory we are on right now. If governments, corporations, and individuals make no further efforts than those registered in 2015, we will go to a warming of at least 3.7 degrees Celsius by 2100. Worse yet, if they do not fulfill even the registered commitments, we can expect warming of 4 or 5 degrees. (See the appendix, page 179.) Be forewarned, this picture is dark. Even though many of the worst-case scenarios might not be realised until the second half of the century, it is clear that by mid-century human misery would be high, biodiversity would be decimated, and that we and our children would live in a world that is constantly deteriorating with no possible recuperation.

The world we must create, limiting warming to no more than 1.5 degrees Celsius.[16] We cannot turn back the clock on past emissions. However, even at this late stage, we can strive for and achieve a better world in which nature and the human family will not only survive but thrive together. Scientists have been extremely clear that the 1.5-degree-Celsius-warmer scenario is still attainable but that the window is rapidly closing. To have at least a 50 per cent chance of success (which in itself is an unacceptably high level of risk), we must cut global emissions to half their current levels by 2030, half again by 2040, and finally to net-zero by 2050 at the very latest.[17] A change of this magnitude would require major transformations in almost every area of life and work, from massive reforestation to new agricultural practices; from the cessation of coal production by 2020 and of oil and gas extraction soon thereafter, to the abandonment of fossil fuels and even the internal combustion engine.

Precisely what we need to do is detailed later in the book but, for now, we have to wake up to the fact that we can choose our future and collectively create it. Our collective responsibility is to ensure that a better future is not only possible but probable, and then not only probable but foreseeable.

The great baseball player Yogi Berra famously said that predictions are hard to make, especially about the future. In constructing these scenarios, we are aware that making predictions about the world in 30 years' time is to some degree an imaginative enterprise. However, everything we set out in these scenarios is predicted or expected by the best science.[18] Indeed, much of what science has foretold is already happening. Read each scenario not as a prediction of the future but as a warning of what may come and what we still have a chance to change.

CHAPTER TWO

The World We Are Creating

It is 2050. Beyond the emissions reductions registered in 2015, no further efforts were made to control emissions. We are heading for a world that will be more than 3 degrees Celsius warmer by 2100.

The first thing that hits you is the air.

In many places around the world, the air is hot, heavy and, depending on the day, clogged with particulate pollution. Your eyes often water. Your cough never seems to disappear. You think about some countries in Asia, where out of consideration sick people used to wear white masks to protect others from airborne infection. Now you often wear a mask to protect yourself from air pollution. You can no longer simply walk out your front door and breathe fresh air: there might not be any. Instead, before opening doors or windows in the morning, you check your phone to see what the air quality will be. Everything might look fine – sunny and clear – but you know better. When storms and

heatwaves overlap and cluster, the air pollution and intensified surface ozone levels can make it dangerous to go outside
without a specially designed face mask (which only some
can afford).[1]

Southeast Asia and Central Africa lose more lives to filthy
air than do Europe or the United States.[2] Fewer people work
outdoors, and even indoors the air can taste slightly acidic,
sometimes making you feel nauseated. The last coal furnaces
closed ten years ago, but that hasn't made much difference
in air quality around the world because you are still breathing dangerous exhaust fumes from millions of cars and buses
everywhere. Some countries have experimented with seeding rain clouds – the process of artificially inducing rain
– hoping to wash pollution out of the sky, but results are
mixed. Seeding clouds to artificially create more rain is difficult and unreliable, and even the wealthiest countries cannot
achieve consistent results.[3] In Europe and Asia, the practice
has triggered international incidents because even the most
skilled experts can't control where the rain will fall, never
mind that acid rain is deleterious to crops, wreaking havoc
on food supply.[4] As a result, crops are increasingly grown
under cover, a trend that will only increase.[5]

Our world is getting hotter. Over the next two decades,
projections tell us that temperatures in some areas of the
globe will rise even higher, an irreversible development now
utterly beyond our control. Oceans, forests, plants, trees and
soil had for many years absorbed half the carbon dioxide
we spewed out. Now there are few forests left, most of them

either logged or consumed by wildfire, and the permafrost is belching greenhouse gases into an already overburdened atmosphere.[6]

The increasing heat of the Earth is suffocating us, and in five to ten years, vast swathes of the planet will be increasingly inhospitable to humans. We don't know how hospitable the arid regions of Australia, South Africa and the Western United States will be by 2100. No one knows what the future holds for their children and grandchildren: tipping point after tipping point is being reached, casting doubt on the form of future civilisation. Some say that humans will be cast to the winds again, gathering in small tribes, hunkered down and living on whatever patch of land might sustain them.[7]

The passing tipping points have already been painful. First was the vanishing of coral reefs. Some of us still remember diving amid majestic coral reefs, brimming with multicoloured fish of all shapes and sizes. Corals are now almost gone. The Great Barrier Reef in Australia is the largest aquatic cemetery in the world. Efforts have been made to grow artificial corals farther north and south from the equator where the water is a bit cooler, but these efforts have largely failed, and marine life has not returned. Soon there will be no reefs anywhere – it is only a matter of a few years before the last 10 per cent dies off.[8]

The second tipping point was the melting of the ice sheets in the Arctic. There is no summer Arctic sea ice anymore because warming is worse at the poles – between 6 and

8 degrees higher than other areas. The melting happened silently in that cold place far north of most of the inhabited world, but its effects were soon noticed. The Great Melting was an accelerant of further global warming. The white ice used to reflect the sun's heat, but now it's gone, so the dark sea water absorbs more heat, expanding the mass of water and pushing sea levels even higher.[9]

More moisture in the air and higher sea surface temperatures have caused a surge in extreme hurricanes and tropical storms. Recently, coastal cities in Bangladesh, Mexico, the United States and elsewhere have suffered brutal infrastructure destruction and extreme flooding, killing many thousands and displacing millions. This happens with increasing frequency now.[10] Every day, because of rising water levels, some part of the world must evacuate to higher ground. Every day the news shows images of mothers with babies strapped to their backs, wading through floodwaters, and homes ripped apart by vicious currents that resemble mountain rivers. News stories tell of people living in houses with water up to their ankles because they have nowhere else to go, their children coughing and wheezing because of the mould growing in their beds, insurance companies declaring bankruptcy leaving survivors without resources to rebuild their lives. Contaminated water supplies, sea salt intrusions, and agricultural run-off are the order of the day. Because multiple disasters are often happening simultaneously, it can take weeks or even months for basic food and water relief to reach areas pummelled by extreme floods. Diseases such as malaria, dengue, cholera, respiratory illnesses, and malnutrition are rampant.[11]

Now all eyes are on the western Antarctic ice sheet, which crossed the point of no return in the early 2020's.[12] If it ever did disappear, it would release a deluge of fresh water into the oceans, potentially raising sea levels by over five metres. If that were to happen, cities like Miami, Shanghai and Dhaka would be uninhabitable – ghostly Atlantises dotting the coasts of each continent, their skyscrapers jutting out of the water, their people evacuated or dead.

Those around the world who chose to remain on the coast because it had always been their home have more to deal with than rising water and floods – they must now witness the demise of a way of life based on fishing. As oceans have absorbed carbon dioxide, the water has become more acidic, and the pH levels are now so hostile to marine life that all but a few countries have banned fishing, even in international waters.[13] Many people insist that the few fish that are left should be enjoyed while they last – an argument, hard to fault in many parts of the world, that applies to so much that is vanishing.

As devastating as rising oceans have been, droughts and heatwaves inland have created a special hell. Vast regions have succumbed to severe aridification, sometimes followed by desertification,[14] and wildlife there has become a distant memory.[15] These places can barely support human life; their aquifers have dried up. Cities like Marrakech and Volgograd are on the verge of becoming deserts. Hong Kong,

Barcelona, Abu Dhabi and many others have been desalin-
ating seawater for years, desperately trying to keep up with
the constant wave of immigration from areas that have gone
completely dry.

Extreme heat is on the march. If you live in Paris, you
endure summer temperatures that regularly rise to 44
degrees Celsius (111 degrees Fahrenheit). This is no longer
the headline-grabbing event it would have been 30 years ago.
Everyone stays inside, drinks water and dreams of air con-
ditioning. You lie on your couch, a cold wet towel over your
face, and try to rest without dwelling on the poor farmers
on the outskirts of town who, despite recurrent droughts
and wildfires, are still trying to grow grapes, olives or soy –
luxuries for the rich, not for you.

You try not to think about the 2 billion people who live
in the hottest parts of the world, where, for upwards of 45
days per year, temperatures skyrocket to 60 degrees Celsius
(140 degrees Fahrenheit) – a point at which the human body
cannot be outside for longer than about six hours because
it loses the ability to cool itself down. Places such as central
India are becoming increasingly challenging to inhabit. For
a while people tried to carry on, but when you can't work
outside, when you can fall asleep only at 4am for a couple
of hours because that's the coolest part of the day, there's not
much you can do but leave. Mass migrations to less hot rural
areas are beset by a host of refugee problems, civil unrest, and
bloodshed over diminished water availability.[16]

Inland glaciers around the world are quickly disappearing.

The millions who depended on the Himalayan, Alpine and Andean glaciers to regulate water availability throughout the year are in a state of constant emergency: there is little snow turning to ice atop mountains in the winter, so there is no more gradual melting for the spring and summer. Now there are either torrential rains leading to flooding or prolonged droughts. The most vulnerable communities with the least resources have already seen what can ensue when water is scarce: sectarian violence, mass migration and death.

Even in some parts of the United States, there are fiery conflicts over water, battles between the rich who are willing to pay for as much water as they want and everyone else demanding equal access to the life-enabling resource. The taps in nearly all public facilities are locked, and those in restrooms are coin operated. At the federal level, Congress is in an uproar over water redistribution: states with less water demand what they see as their fair share from states that have more. Government leaders have been stymied on the issue for years, and with every passing month the Colorado River and the Rio Grande shrink further.[17] Looming on the horizon are conflicts with Mexico, no longer able to guarantee deliveries of water from the depleted Rio Conchos and Rio Grande.[18] Similar disputes have arisen in Peru, China, Russia and many other countries.

Food production swings wildly from month to month, season to season, depending on where you live. More people are starving than ever before. Climate zones have shifted, so some new areas have become available for agriculture

(Alaska, the Arctic),[19] while others have dried up (Mexico, California). Still others are unstable because of the extreme heat, never mind flooding, wildfire, and tornadoes. This makes the food supply in general highly unpredictable. One thing hasn't changed, though – if you have money, you have access. Global trade has slowed as countries such as China stop exporting and seek to hold on to their own resources. Disasters and wars rage, choking off trade routes. The tyranny of supply and demand is now unforgiving; because of its increasing scarcity, food can now be wildly expensive. Income inequality has always existed, but it has never been this stark or this dangerous.

Entire regions suffer from epidemics of stunting and malnutrition. Reproduction has slowed overall, but most acutely in those countries where food scarcity is dire. Infant mortality has rocketed, and international aid has proven to be politically impossible to defend in light of mass poverty. Countries with enough food are resolute about holding on to it.

In some places, the inability to gain access to such basics as wheat, rice or sorghum has led to economic collapse and civil unrest more quickly than even the most pessimistic experts had previously imagined. Scientists tried to develop varieties of staples that could stand up to drought, temperature fluctuations, and salt, but there was only so much we could do. Now there simply aren't enough resilient varieties to feed the population. As a result, food riots, coups, and civil wars are throwing the world's most vulnerable from the frying pan into the fire. As developed countries seek to seal their borders from mass migration, they too feel the consequences. Stock markets are crashing, currencies are wildly fluctuating, and the European Union has disbanded.[20]

As committed as nations are to keeping wealth and resources within their borders, they're determined to keep people out. Most countries' armies are now just highly militarised border patrols. Lockdown is the goal, but it hasn't been a total success. Desperate people will always find a way. Some countries have been better global Good Samaritans than others, but even they have now effectively shut their borders, their wallets and their eyes.[21]

Ever since the equatorial belt started to become difficult to inhabit, an unending stream of migrants has been moving north from Central America, towards Mexico and the United States. Others are moving south toward the tips of Chile and Argentina. The same scenes are playing out across Europe and Asia. Enormous political pressure is being placed on northern and southern countries to either welcome migrants or keep them out. Some countries are letting people in, but only under conditions approaching indentured servitude. It will be years before the stranded migrants are able to find asylum or settle into new refugee cities that have formed along the borders.

Even if you live in areas with more temperate climates such as Canada and Scandinavia, you are still extremely vulnerable. Severe tornadoes, flash floods, wildfires, mudslides and blizzards are often in the back of your mind. Depending on where you live, you have a fully stocked storm cellar, an emergency go-bag in your car or a 6-foot fire moat around your house. People are glued to oncoming weather reports. Only the foolhardy shut their phones off at night. If an emergency hits, you may only have minutes to respond. The alert systems set up by the government are basic and subject to glitches and irregularities depending on access to technology. The rich, who subscribe to private, reliable satellite-based alert systems, sleep better.

The weather is unavoidable, but lately the news about what's going on at the borders has become too much for most people to endure. Because of the alarming spike in suicides, and under increasing pressure from public health officials, news organisations have decreased the number of stories devoted to genocide, slave trading, and new mutant virus outbreaks and deadly pandemics. You can no longer trust the news. Social media, long the grim source of live feeds and disaster reporting, is brimming with conspiracy theories and doctored videos. Overall, the news has taken a strange, seemingly controlled turn towards distorting reality and spinning a falsely positive narrative.

Those living within stable countries may be physically safe, yes, but the psychological toll is mounting. With each new tipping point passed, they feel hope slipping away. There is no chance of stopping the runaway warming of our planet, and no doubt we are slowly but surely heading towards some kind of collapse. And not just because it's too hot. Melting permafrost is also releasing ancient microbes that today's humans have never been exposed to – and as a result have no resistance to.[22] Diseases spread by mosquitoes and ticks are rampant as these species flourish in the changed climate, spreading to previously safe parts of the planet, increasingly overwhelming us. It's not only diseases spread by insects that feed on our blood that we are struggling against. As deforestation and intensive agriculture encroaches further into our remaining wild spaces, we're battling against constantly emerging zoonotic diseases that threaten to make the COVID-19 pandemic in the early 2020's look mild. Worse still, the public health crisis of antibiotic and vaccine resistance has only intensified as the population has grown denser in inhabitable areas and temperatures continue to rise.[23]

The demise of the human species is being discussed more and more. For many, the only uncertainty is how long we'll last, how many more generations will see the light of day. Suicides are the most obvious manifestation of the prevailing despair, but there are other indications: a sense of bottomless loss, unbearable guilt and fierce resentment at previous generations who didn't do what was necessary to ward off this unstoppable calamity.

CHAPTER THREE

The World We Must Create

It is 2050. We have been successful at halving emissions every decade since 2020. We are heading for a world that will be no more than 1.5 degrees Celsius warmer by 2100.

In most places in the world, the air is moist and fresh, even in cities. It feels a lot like walking through a forest, and very likely this is exactly what you are doing. The air is cleaner than it has been since before the Industrial Revolution.

We have trees to thank for that. They are everywhere.[1]

It wasn't the single solution we required, but the proliferation of trees bought us the time we needed to vanquish carbon emissions. Corporate donations and public money funded the biggest tree-planting campaign in history. When we started, it was purely practical, a tactic to combat climate change by relocating the carbon: the trees took carbon dioxide out of the air, released oxygen, and put the carbon back where it belongs, in the soil. This, of course, helped to diminish climate change, but the benefits were even greater. On every sensory level, the ambient feeling of living on what has again become a green planet has been transformative, especially in cities. Cities have never been better places to live. With many more trees and far fewer cars, it has been possible to reclaim whole streets for urban agriculture and for children's play. Every vacant lot, every grimy unused alley, has been repurposed and turned into a shady grove. Every rooftop has been converted to either a vegetable or a floral garden. Windowless buildings that were once scrawled with graffiti are instead carpeted with verdant vines.

The greening movement in Spain had begun as an effort to combat rising temperatures. Because of Madrid's latitude, it is one of the driest cities in Europe. And even though the city now has a grip on its emissions, it was previously at risk of desertification. Because of the 'heat island' effect of cities – buildings trap warmth and dark, paved surfaces absorb heat from the sun – Madrid, home to more than 6 million people, was several degrees warmer than the countryside just a few miles away. In addition, air pollution was leading to a rising incidence of premature births,[2] and a spike in deaths was linked to cardiovascular and respiratory illnesses. With a health care system already strained by the arrival of subtropical diseases like dengue fever and malaria, government officials and citizens rallied. Madrid made dramatic efforts to reduce the number of vehicles and create a 'green envelope' around the city to help with cooling, oxygenating and filtering pollution. Plazas were repaved with porous material to capture rainwater, all black roofs were painted white and plants were omnipresent. The plants cut noise, released oxygen, insulated south-facing walls, shaded pavements and released water vapour into the air. The massive effort was a huge success and was replicated all over the world. Madrid's economy boomed as its expertise put it on the cutting edge of a new industry.

Most cities found that lower temperatures raised the standard of living. There are still slums, but the trees, largely responsible for countering the temperature rise in most places, have made things far more bearable for all.

Reimagining and restructuring cities was crucial to solving the climate challenge puzzle. But further steps had to be taken, which meant that global rewilding efforts had to reach well beyond the cities. The forest cover worldwide is now 50 per cent,[3] and agriculture has evolved to become more tree-based. The result is that many countries are unrecognisable, in a good way. No one seems to miss wide-open plains or monocultures. Now we have shady groves of nut and fruit orchards, timberland interspersed with grazing, parkland areas that spread for miles, new havens for our regenerated population of pollinators.[4]

Luckily for the 75 per cent of the population who live in cities, new electric railways crisscross interior landscapes. In the United States, high-speed rail networks on the East and West coasts have replaced the vast majority of domestic flights, with East Coast connectors to Atlanta and Chicago. Because flight speeds have slowed down to gain fuel efficiency, passenger bullet trains make some journeys even faster and with no emissions whatsoever.[5] The US Train Initiative was a monumental public project that sparked the economy for a decade. Replacing miles and miles of interstate highways with a new transportation system created millions of jobs – for train technology experts, engineers, and construction workers who designed and built raised rail tracks to circumvent floodplains. This massive effort helped to re-educate and retrain many of those displaced by the dying fossil fuel economy. It also introduced a new generation of workers to the excitement and innovation of the new climate economy.

Running parallel to this mega public works effort was an increasingly confident race to harness the power of renewable sources of energy. A major part of the shift to net-zero emissions was a focus on electricity; achieving the goal required not only an overhaul of existing infrastructure but also a structural shift. In some ways, breaking up grids and decentralising power proved easy. We no longer burn fossil fuels. There is some nuclear energy in those countries that can afford the expensive technology,[6] but most of our energy now comes from renewable sources like wind, solar, geothermal, and hydro. All homes and buildings produce their own electricity – every available surface is covered with solar paint that contains millions of nanoparticles, which harvest energy from the sunlight,[7] and every windy spot has a wind turbine. If you live on a particularly sunny or windy hill, your house might harvest more energy than it can use, in which case the energy will simply flow back to the smart grid. Because there is no combustion cost, energy is basically free. It is also more abundant and more efficiently used than ever.

Smart tech prevents unnecessary energy consumption, as artificial intelligence units switch off appliances and machines when not in use. The efficiency of the system means that, with a few exceptions, our quality of life has not suffered. In many respects, it has improved.

For the developed world, the wide-ranging transition to renewable energy was at times uncomfortable, as it often involved retrofitting old infrastructure and doing things in new ways. But for the developing world, it was the dawn of a new era. Most of the infrastructure that it needed for economic growth and poverty alleviation was built according to the new standards: low-carbon emissions and high resilience. In remote areas, the billion people who had no electricity at the start of the twenty-first century now have energy generated by their own rooftop solar modules or by wind-powered minigrids in their communities. This new access opened the door to so much more. Entire populations have leaped forwards with improved sanitation, education, and health care. People who had struggled to get clean water can now provide it to their families. Children can study at night. Remote health clinics can operate effectively.

Homes and buildings all over the world are becoming self-sustaining far beyond their electrical needs. For example, all buildings now collect rainwater and manage their own water use. Renewable sources of electricity made possible localised desalination, which means clean drinking water can now be produced on demand anywhere in the world. We also use it to irrigate hydroponic gardens, flush toilets, and shower.[8] Overall, we've successfully rebuilt, reorganised, and restructured our lives to live in a more localised way. Although energy prices have dropped dramatically, we are choosing local life over long commutes. Due to greater connectivity, many people work from home, allowing for more flexibility and more time to call their own.

We are making communities stronger. As a child, you might have seen your neighbours only in passing. But now, to make things cheaper, cleaner and more sustainable, your orientation in every part of your life is more local. Things that used to be done individually are now done communally – growing vegetables, capturing rainwater and composting. Resources and responsibilities are shared now. At first you resisted this *togetherness* – you were used to doing things individually and in the privacy of your own home. But pretty quickly the camaraderie and unexpected new network of support started to feel good, something to be prized. For most people, the new way has turned out to be a better recipe for happiness.

Food production and procurement are a big part of the communal effort. When it became clear we needed to revolutionise industrialised farming, we transitioned quickly to regenerative farming practices – mixing perennial crops, sustainable grazing, and improved crop rotation on large-scale farms, with increased community reliance on small farms.[9] Instead of going to a big supermarket for food flown in from hundreds, if not thousands, of miles away, you buy most of your food from small local farmers and producers. Buildings, neighbourhoods and even large extended families form a food purchase group, which is how most people buy their food now. As a unit they sign up for a weekly drop-off, then distribute the food among the group members. Distribution, coordination, and management are everyone's responsibility, which means you might be partnered with a downstairs neighbour for distribution one week and your upstairs neighbour the next.

While this community approach to food production makes things more sustainable, food is still expensive, consuming up to 30 per cent of household budgets, which is why growing your own is such a necessity.[10] In community gardens, on rooftops, at schools and even hanging from vertical gardens on balconies, food sometimes seems to be growing everywhere.

We've come to realise, by growing our own, that food is expensive because it *should* be expensive – it takes valuable resources to grow it, after all. Water. Soil. Sweat. Time.[11] For that reason, the most resource-depleting foods of all – animal protein and dairy products – have practically disappeared from our diets.[12] But the plant-based replacements are so good that most of us don't notice the absence of meat and dairy. Most young children cannot believe we used to kill any animals for food. Fish is still available, but it is farmed and yields are better managed by improved technology.[13]

We make smarter choices about bad foods, which have

become an ever-diminishing part of our diets. Government taxes on processed meats, sugars, and fatty foods helped us reduce the carbon emissions from farming. The biggest boon of all was to our collective health. Thanks to reduced cancers, heart attacks, and strokes, people are living longer, and health services around the world cost less and less. In fact, a huge portion of the costs of combating climate change were recuperated by government savings on public health.[14]

Along with outrageous spending on health care, petrol and diesel cars are also anachronisms. Most countries banned their manufacture in 2030,[15] but it took another 15 years to get internal combustion engines off the road completely. Now they are seen only in transport museums or at special rallies where classic car owners pay an offset fee to drive a few short miles around the track. And of course, they are all hauled in on the backs of huge electric trucks.

When it came to making the switch, some countries were already ahead of the curve. Technology-driven countries such as Norway and bicycle-friendly nations like the Netherlands managed to impose a moratorium on cars much earlier. Unsurprisingly, the United States had the hardest time of all. First, it restricted their sale, and then it banned them from certain parts of cities – Ultra Low Emission Zones.[16] Then came the breakthrough in the battery storage capacity of electric vehicles,[17] the cost reductions that came from finding alternative materials for manufacture, and finally the complete overhaul of the charging and parking infrastructure.[18] This allowed people easier access to cheap power for their electric vehicles. Even better, car batteries are now bidirectionally connected with the electric grid, so they can either charge from the grid or provide power to the grid when they aren't being driven. This helps back up the smart grid that is running on renewable energy.

The ubiquity and ease of electric vehicles were alluring, but satisfaction of our appetite for speed finally did the trick.[19] Supposedly, to stop a bad habit you have to replace it with one that is more salubrious or at least as enjoyable. At first, China dominated the manufacture of electric vehicles, but soon US companies started making vehicles that were more desirable than ever before. Even some classic cars got an upgrade, switching from combustion to electric engines that could go from zero to 60 mph in 3.5 seconds.[20] What's strange is that it took us so long to realise that the electric motor is simply a better way of powering vehicles. It gives you more torque, more speed when you need it, and the ability to recapture energy when you brake, and it requires dramatically less maintenance.

As people from rural areas moved to the cities, they had less need even for electric vehicles.[21] In cities it's now easy to get around – transportation is frictionless. When you take the electric train, you don't have to fumble around for a metro card or wait in line to pay – the system tracks your location, so it knows where you got on and where you got off, and it deducts money from your account accordingly. We also share cars without thinking twice. In fact, regulating and ensuring the safety of driverless ride sharing was the biggest transportation hurdle for cities to overcome. The goal has been to eliminate private ownership of vehicles by 2050 in major metropolitan areas.[22] We're not quite there yet, but we're making progress.

We have also reduced land transport needs. Three-dimensional (3D) printers are readily available, cutting down on what people need to purchase away from home.[23] Drones organised along aerial corridors are now delivering packages, further reducing the need for vehicles.[24] Thus we are currently narrowing roads, eliminating parking spaces, and investing in urban planning projects that make it easier to walk and bike in the city. Parking garages are used only for ride sharing, electric vehicle charging and storage – those ugly concrete stacking systems and edifices of yore are now enveloped in green. Cities now seem designed for the coexistence of people and nature.

International air travel has been transformed. Biofuels have replaced jet fuel. Communications technology has advanced so much that we can participate virtually in meetings anywhere in the world without travelling. Air travel still exists, but it is used more sparingly and is extremely costly. Because work is now increasingly decentralised and can often be done from anywhere, people save and plan for 'slow-cations' – international trips that last weeks or months instead of days. If you live in the United States and want to visit Europe, you might plan to stay there for several months or more, working your way across the continent using local, zero-emissions transportation.[25]

While we may have successfully reduced carbon emissions, we're still dealing with the after-effects of record levels of carbon dioxide in the atmosphere. The long-living greenhouse gases have nowhere to go other than the already-loaded atmosphere, so they are still causing increasingly extreme weather – though it's less extreme than it would have been had we continued to burn fossil fuels. Glaciers and Arctic ice are still melting, and the sea is still rising. Severe droughts and desertification are occurring in the western United States, the Mediterranean, and parts of China. Ongoing extreme weather and resource degradation continue to multiply existing disparities in income, public health, food security and water availability. But now governments have recognised climate change factors for the threat multipliers that they are. That awareness allows us to predict downstream problems and head them off before they become humanitarian crises.[26] So while many people remain at risk every day, the situation is not as drastic or chaotic as it might have been.[27] Economies in developing nations are strong, and unexpected global coalitions have formed with a renewed sense of trust. Now when a population is in need of aid, the political will and resources are available to meet that need.

The ongoing refugee situation has been escalating for decades, and it is still a major source of strife and discord. But around 15 years ago, we stopped calling it a crisis. Countries agreed on guidelines for managing refugee influxes – how to smoothly assimilate populations, how to distribute aid and resources, and how to share the tasks within particular regions. These agreements work well most of the time, but things get thrown off balance occasionally when a country flirts with fascism for an election cycle or two.

Technology and business sectors stepped up, too, seizing the opportunity of government contracts to provide large-scale solutions for distributing food and providing shelter for the newly displaced. One company invented a giant robot that could autonomously build a four-person dwelling within days.[28] Automation and 3D printing have made it possible to quickly and affordably construct high-quality housing for refugees. The private sector has innovated with water transportation technology and sanitation solutions. Fewer tent cities and housing shortages have led to less cholera.

Everyone understands that we are all in this together. A disaster that occurs in one country is likely to occur in another in only a matter of years. It took us a while to realise that if we worked out how to save the Pacific Islands from rising sea levels this year, then we might find a way to save Rotterdam in another five years. It is in the interest of every country to bring all its resources to bear on problems across the world. For one thing, creating innovative solutions to climate challenges and beta-testing them years ahead of using them is just plain smart. For another, we're nurturing goodwill; when we need help, we know we will be able to count on others to step up.

The zeitgeist has shifted profoundly. How we feel about

the world has changed, deeply. And, unexpectedly, so has how we feel about one another.

When the alarm bells rang in 2020, thanks in large part to the youth movement, we realised that we suffered from too much consumption, competition, and greedy self-interest. Our commitment to these values and our drive for profit and status had led us to steamroll our environment. As a species we were out of control, and the result was the near collapse of our world. We could no longer avoid seeing on a tangible, geophysical level that when you spurn regeneration, collaboration, and community, the consequence is impending devastation.

Extricating ourselves from self-destruction would have been impossible if we hadn't changed our mindset and our priorities, if we hadn't realised that doing what is good for humanity goes hand in hand with doing what is good for the Earth. The most fundamental change was that collectively – as citizens, corporations and governments – we began adhering to a new bottom line: 'Is it good for humanity whether profit is made or not?'

The climate change crisis of the beginning of the century jolted us out of our stupor. As we worked to rebuild and care for our environment, it was only natural that we also turned to each other with greater care and concern. We realised that the perpetuation of our species was about far more than saving ourselves from extreme weather. It was about being good stewards of the land *and* of one another. When we began the fight for the fate of humanity, we were thinking only about the species' survival but, at some point, we understood that it was as much about the fate of *our humanity*. We emerged from the climate crisis as more mature members of the community of life, capable of not only restoring ecosystems but also of unfolding our dormant potentials of human strength and discernment. Humanity was only ever as doomed as it believed itself to be. Vanquishing that belief was our true legacy.

PART II
THREE MINDSETS

CHAPTER FOUR

Who We Choose to Be

Our future is unwritten. It will be shaped by who we choose to be now.

As we learned during our stewardship of the Paris Agreement, if you do not control the complex landscape of a challenge (and you rarely do), the most powerful thing you can do is change how you behave in that landscape, using yourself as a catalyst for overall change. All too often in the face of a task, we move quickly to 'doing' without first reflecting on 'being' – what *we* personally bring to the task, as well as what others might. And the most important thing we can bring is our state of mind.

Mahatma Gandhi reminds us to be the change we want to see. The actions we pursue are largely defined by the mindset we cultivate in advance of the doing. Faced with an urgent task, it may feel counterintuitive to first look inside ourselves, but it is essential.

Attempting change while we are informed by the same state of mind that has been predominant in the past will lead to insufficient incremental advances. In order to open the space for transformation, we have to change how we think and fundamentally who we perceive ourselves to be. After all, if what's at stake is nothing less than the quality of human life for centuries to come, it is worth digging down to the roots of who we understand ourselves to be.

Paradoxically, systemic change is a deeply personal endeavour. Our social and economic structures are a product of our way of thinking.

For example, our economy is based on the belief that we can extract resources boundlessly, use them inefficiently, and discard them wantonly, drawing from the planet more than it can regenerate and polluting more than we can clean up. Over time we've developed a deeply exploitative ethos as the basis of our actions.

This no longer works.

Natural scientists have provided ample evidence that we have reached several planetary boundaries, beyond which Earth's biosystems cannot sustain life. Soon there will be little left to extract and exploit. Concerned social scientists are clear on what we need to do: we must move towards a regenerative economy, an economy that operates in harmony with nature, repurposing used resources, minimising waste, and replenishing depleted resources. We must return to the innate wisdom of nature herself, the ultimate regenerator and recycler of all resources.

Less understood but just as important is the fact that we have reached the limits of our individualistic competitive approach. For a long time, Western societies have tended to prize self-interest over the well-being of the whole. We need to enlarge our understanding of ourselves and our relationships with others, and certainly with the natural systems that enable human life on Earth.

Our current crisis requires a total shift in our thinking. To survive and thrive, we must understand ourselves to be inextricably connected to all of nature. We need to cultivate a deep and abiding sense of stewardship. This transformation begins with the individual. Who we are and how we show up in the world defines how we work with others, how we interact with our surroundings and ultimately the future we co-create.

We believe three mindsets are fundamental to us all in our pursuit to co-create a better world. With intentional provocation, we call them Stubborn Optimism, Endless Abundance and Radical Regeneration. These mindsets are not new. We can find shining examples in famous historical figures, but the future we want requires that they be prevalent among us all. These qualities of being are innate human capacities (individual and collective), values that can be called forth, nurtured and developed in the crucible of daily practice.

A shift in consciousness may sound grandiose to some, insufficient to others. But we live at a time of growing awareness of the intimate connections between the outer and inner worlds. As author Joanna Macy has pointed out, 'In the past changing the self and changing the world were regarded as separate endeavours and viewed in either-or terms. That is no longer the case.'[1] Scientific understanding and spiritual insights are converging on the reality of human-nature interconnectedness.

The transformative power of the three mindsets lies not only in themselves but also in the direction each one provides. Attached as we are to many forms of status quo in our lives (relationships, job, home, etc.), we often delude ourselves that they are permanent. But the fact is, nothing is permanent; everything is always changing, no matter how much we insist on standing still, hanging on to fleeting moments. And making desired change always demands going in an intentional direction.

Our new intentional direction must move us beyond defeatism to optimism, beyond extraction towards regeneration, beyond linear towards circular economies, beyond individual benefit towards the common good, beyond short-term thinking towards long-term thinking and acting. By cultivating the three mindsets, we give clearer, stronger direction to our lives and to our world, setting the necessary foundation for us to collectively co-create the world we want.

CHAPTER FIVE

Stubborn Optimism

Twenty-five hundred years ago, Siddhartha Gautama, the man who became known as the Buddha, understood optimism. He said many times that a brightness of mind was both the final goal of the path of enlightenment and also the first step. A bright mind is how you proceed. Without it, you can't make progress.

The Buddha also understood that we are not subject to our attitudes in a passive way but are active participants in creating them. Neuroscience has now confirmed this. It does not matter if our natural tendency is to see things with optimism or with pessimism. At this point in history we have a responsibility to do what is necessary, and for most of us that will involve some deliberate reprogramming of our minds.

Psychological research has shown that attitudes can be transformed by first identifying our thought patterns, then deliberately cultivating a more constructive approach. The practice involves becoming aware of these patterns, drawing out the unconscious assumptions, and challenging them when they don't serve you.[1]

It's not complicated, but neither is it easy. Essentially, we all have inbuilt reactions to adverse things that happen around us. From the latest alarming report on climate change to missing the bus, we have a learned response to all phenomena that we encounter in life, and those learned reactions dictate how we respond to a particular situation. When it comes to climate change, the vast majority of us have a learned reaction of helplessness. We see the direction the world is headed, and we throw up our hands. Yes, we think, it's terrible, but it's so complex and so big and so overwhelming. We can't do anything to stop it.

This learned reaction is not only untrue, it's become fundamentally irresponsible. If you want to help address climate change, you have to teach yourself a different response.

You can do it. You can switch your focus, and you will be stunned by the impact such a shift can create. You don't need to have all the answers, and you certainly don't need to hide from the truth, nor should you. When you are faced with the hard realities, look at them with clarity, but also know that you are incredibly lucky to be alive at a time when you can make a transformative difference to the future of all life on Earth.

You are not powerless. In fact, your every action is suffused with meaning, and you are part of the greatest chapter of human achievement in history. Make this your mental mantra. Take notice of how your mind tries to insist on your helplessness in the face of the challenge and refuses to accept it. Notice it, and refute it. It will not take long for your thought patterns to change.

When your mind tells you that it is too late to make a difference, remember that every fraction of a degree of extra warming makes a big difference, and therefore any reduction in emissions lessens the burden on the future.

When your mind tells you that this is all too depressing to deal with and that it is better to focus on the things you can directly affect, remind yourself that mobilising for big generational challenges can be thrilling and can imbue your life with meaning and connection.

When your mind tells you that it will be impossible for the world to lighten its dependence on fossil fuels, remember that already more than 50 per cent of the energy in the UK comes from clean power,[2] that Costa Rica is 100 per cent clean,[3] and that California has a plan to get to 100 per cent clean, including cars and trucks, by the time today's toddlers have finished college.[4]

When your mind tells you that the problem is the broken political system and we can't fix that so there is no point in doing anything, remind yourself that political systems are still responsive to the views of people, and that throughout history people have successfully overcome extraordinary odds to achieve political change.

And when your mind tells you that you are just one person, too small to make a difference, so why bother, you can remind yourself that tipping points are non-linear. We don't know what is going to make the difference, but we know that in the end systems do shift and all the little actions add up to a new world. Every time you make an individual choice to be a responsible custodian of this beautiful Earth, you contribute to major transformations.

You may not be religious or spiritually inclined, but consider the lot of the stonemason in medieval Europe building one of the great cathedrals. He could have chosen to throw down his tools because he was not going to personally finish the entire cathedral. Instead, he worked patiently and carefully on his one piece, knowing he was part of a great collective endeavour that would lift the hearts and minds of generations. That is optimism, and cultivating it will not only be a crucial step to advancing our human story, it will also improve your life today.

Václav Havel aptly described optimism as 'a state of mind, not a state of the world.'[5] Three characteristics are generally agreed upon as essential to making this mindset transformative: the intention to see beyond the immediate horizon, the comfort with uncertainty about the final outcome, and the commitment that is fostered by that mindset.

To be optimistic, you must acknowledge the bad news that is all too readily available in scientific reports, your newsfeed, your Twitter account, and kitchen table conversations bemoaning our current state of affairs. More difficult, but necessary for any degree of change to take place, is to recognise the adversities and still be able to see that a different future is not only possible but is already tiptoeing into our daily lives. Without denying the bad news, you must make a point of focusing on all the good news regarding climate change, such as the constantly dropping prices of renewables, an increasing number of countries taking on net-zero-emissions targets by 2050 or before, the multiple cities banning internal combustion vehicles and the rising levels of capital shifting from the old to the new economy. None of this is happening yet at the necessary scale, but it is happening. Optimism is about being able to intentionally identify and prescribe the desired future so as to actively pull it closer.

It is always easier to cling to certainty than it is to work for something because it is right and good, regardless of whether it currently stands a decent chance of success. All the measures to address climate change still require further maturation; none guarantee ultimate success. We don't know which renewables, if any, will predominate, or which are more likely to scale quickly. Problems with the batteries of electric vehicles (weight, cost, recycling) must still be solved, and charging networks still require substantial expansion to succeed. Financial instruments must more effectively manage the risks of new technologies. Market models that shift us from single ownership of homes and cars to shared ownership must gather steam and make peace with regulation.

When you look at the future broadly instead of narrowly, you see that you must take these uncertainties in your stride, or you will stay stuck in the knowns of the past. You have to be willing to risk mistakes, delays and disappointments, or you will be at the mercy of only the tried and true, to your ultimate peril.

This mindset is all the more important once you realise that the habits, practices and technologies of the past will lead us only to ecological demise and human suffering. Viewing our reality with optimism means recognising that another future is possible, not promised. In the face of climate change, we all have to be optimistic, not because success is guaranteed but because failure is unthinkable.

Optimism empowers you; it drives your desire to engage, to contribute, to make a difference. It makes you jump out of bed in the morning because you feel challenged and hopeful at the same time. It calls you to that which is emerging and makes you want to be an active part of change. Rebecca Solnit puts it well: 'Hope is an axe you break down doors with in an emergency; … hope should shove you out the door, because it will take everything you have to steer the future away from endless war, from the annihilation of the earth's treasures and the grinding down of the poor and marginal… . To hope is to give yourself to the future, and that commitment to the future makes the present inhabitable.'[6]

In other words, optimism is the force that enables you to create a new reality.

Optimism is not the *result* of achieving a task we have set for ourselves. That is a celebration. Optimism is the necessary *input* to meeting a challenge.

Optimism is about having steadfast confidence in our ability to solve big challenges. It is about making the choice to tenaciously work to make the current reality better.

Optimism is about actively proving, through every

decision and every action, that we are capable of designing a better future.

From the darkness of an Alabama jail, Martin Luther King, Jr., kept calling for the realisation of a deeply held dream, no matter how bleak its prospects. Many others have done the same throughout history: John F. Kennedy refusing to accept that nuclear war was inevitable; Gandhi marching to the ocean to collect forbidden salt.

In all these cases, key people believed that a better world was possible, and they were willing to fight for it. They didn't ignore difficult evidence or present things in a way that wasn't true. Instead they faced reality with a fierce belief that change could happen, however impossible it might have seemed at the moment.

On the road to the Paris Agreement in 2015, we learned just how critical optimism is to transformation. When Christiana took over responsibility for the United Nations' annual rounds of climate negotiations in 2010, it was in the wake of a total collapse of the previous year's negotiations, which had been held in Copenhagen.

Copenhagen was nothing short of a disaster. After years of preparation and two weeks of excruciating around-the-clock negotiations, the only result was a weak, inadequate accord that was politically unacceptable and legally irrelevant. The United States had embarrassingly declared success prematurely. China and India had put up major roadblocks, supported by all developing countries. It had been a free-for-all of political frustration, outrage and disagreement.

It was far from the 'Hopenhagen' the hosts had advertised. In fact, there had been blood.

Claudia Salerno, the Venezuelan representative, had been excluded from the small room where only a few leaders had negotiated behind closed doors. She was so angry and so adamant about getting the floor, she incessantly banged her country's metal nameplate on her desk until her hand was bleeding.

'Do I have to bleed to get your attention?' she screamed at the Danish chairman. 'International agreements cannot be imposed by a small exclusive group. You are endorsing a coup d'état against the United Nations.'

Each sentence was punctuated with the pounding of metal and blood.

If this is what saving the planet looked like, we were all doomed.

Six months later, UN Secretary-General Ban Ki-moon asked Christiana to assume responsibility for the international climate negotiations. There was little hope in his request: pick up the pieces from the political garbage can and make something of them.

No one, from a high-level administrator at the UN to a government delegate to a climate activist working from home, believed that the world had a shot at ever achieving a workable agreement. Everyone thought it was too complicated, too costly and too late anyway.

As a result, one of the toughest challenges Christiana

faced was bringing everyone to believe that an agreement was even possible. Prior to considering the political, technical, and legal parametres of an eventual agreement, she knew she had to dedicate herself to changing the mood on climate. The impossible had to be made possible.

The very first step was to change her own attitude.

As the recently appointed Executive Secretary of the United Nations Convention on Climate Change, Christiana held her first and best-remembered press conference. The new voice of the entire international process, she sat before 40 journalists, gathered in a windowless room in the Maritim Hotel in Bonn, Germany.

After a few anodyne interjections, the most important question was asked: 'Ms. Figueres, do you think a global agreement will ever be possible?'

Without thinking, she blurted, 'Not in my lifetime.'

Christiana had instinctively spoken for the thousands of people who had been in Copenhagen, and for the millions more who followed the proceedings online. Hope was gone, and the pain was deep. Her words expressed the prevailing mood, but they also ripped straight into her own heart. The attitude she had just perpetuated was exactly the problem. If she succumbed to despair, and by extension let this whole political process succumb to it, it would define the quality of life for millions of vulnerable people today and determine the fate of future generations. She couldn't let that happen.

Impossible is not a fact. It is an attitude.

When Christiana walked out of the press conference that day, she knew her primary task: to be a beacon of possibility that would allow everyone to find a way to a solution together. How it would happen she did not know, but she knew with clarity that she had no other option.

Bringing about a complex, large-scale transformation is akin to weaving a tapestry of elaborate design with thousands of people who have never woven anything or even seen the pattern. Almost 200 nations, 500 supporting UN staff members, more than 60 topics under negotiation across five (sometimes intersecting) negotiating tracks, and thousands of participants from all walks of life were involved. Of course, everyone wanted a good future for humanity, but once you dove just one level below that very basic goal, everything else was under constant negotiation, from agreeing on the agenda for one working session, to topics as contentious as how science should be reflected in policy. Predictably, setbacks and obstructions quickly became the norm.

Throughout the whole process, we paid attention to the underlying challenging dynamics, guiding them into a constructive space so that innovative solutions could emerge from the fertile ground of collective participation and wisdom. Careful and well-targeted interventions were repeatedly necessary to ensure forward momentum but could never become overbearing. The intention was to constantly unblock pent-up energy and catalyse the next level of work. Complex dynamic systems can be intimidating if approached from the expectation of control, but they are thrilling if seen as a carefully curated landscape of potential that blossoms as problematic issues find resolution and enrich the commonly agreed ground.

In December 2015, 195 nations adopted the Paris Agreement unanimously, and hundreds of millions of people widely recognised it as a historic achievement. Undoubtedly many factors contributed to this resounding success, as well as thousands of individuals, but the key was the contagious frame of mind that led to collective wisdom and effective decision-making. Everyone who was there at the adoption, and millions of people following online, felt optimistic about the future, but in fact optimism had been the starting point of the journey. It had had to be, or else we would never have reached any agreement.

We need to remember, however, that in the challenging years to come, optimism on its own won't be enough, as it wasn't enough in Paris. What sustained us through the long nights and years of building that initial agreement was a particular brand of optimism that is necessary for the most difficult tasks: stubborn optimism.

Optimism is not soft, it is gritty. Every day brings dark news, and no end of people tell us that the world is going to hell. To take the low road is to succumb. To take the high road is to remain constant in the face of uncertainty. That we may be confronted by barriers galore should not surprise anyone. That we may see worsening climate conditions in the short term should also not surprise us. We have to elect to boldly persevere. With determination and utmost courage, we must conquer the hurdles in order to push forward.

We need both systemic transformation and individual behavioural changes. One without the other will not get us to the necessary scale of change at the necessary pace. We all sit at various points of society: members of families, community leaders, CEOs, policy makers. No matter where you sit, we all can and must exercise that responsibility in favour of the common good. No one is irrelevant.

Particularly in the face of grand human challenges, the only responsible approach we can take is to protect humanity and other forms of life and steer the course of history towards the better. Changing direction at this late hour is entirely possible, but only with a collective intent and optimism that is so robust, we jolt ourselves out of the currently established default path.

The story of the five-year process towards Paris is in many ways like the process we must now unleash. Today, most people believe it is impossible to transform our economy in one decade. But we cannot afford that fatalism; our only option is to turn our full attention to the immediate actions we can undertake to change direction. It starts with our own way of thinking about the challenge, our determined attitude, and our capacity to infect others with the same conviction, no matter how challenging that is. That is stubborn optimism.

The evolution of humanity is a story of adaptive ingenuity to the challenges of the time. We face the greatest challenge of human history. We may be challenged beyond our currently visible capacities, but that only means that we are invited to rise to the next level of our abilities. And we can.

CHAPTER SIX

Endless Abundance

The feeling that we have to compete with others to get what we want, or what we think we need, runs deep in each of us. Most of us have grown up under the stifling influence of the zero-sum paradigm, the notion that if one person wins, another one *has* to lose. (One person's gain has to be 'balanced' by another's loss in order for the sum of all gains and losses to be zero.) The zero-sum paradigm has baked competition into our worldview. Without competition, we could not have achieved many of the great economic and social advances we have made over the centuries. And we will still need a healthy competitive edge to develop the new technologies that will help us address climate change. But when we allow competition to become the dominant feature of our decision-making, we lose our grounding and start to see scarcity in places it may not even exist.

Few of us haven't felt that rush of urgency and determination to get ahead of the crowd for a seat on the train or bus. It's a feeling so ubiquitous that in some countries transportation companies have announcements reminding us to let passengers off the bus or train before attempting to board. But the drive to compete for a seat is sometimes so strong, the announcements cannot prevent people from pushing in first to claim their spot.

The frenzy that dominates in these scenarios doesn't begin with our competitive impulse. It starts with the deeply ingrained *perception* of scarcity – the view that there is a limited amount of something regardless of what the reality may be. We are convinced that there is only one good seat, so we want to secure it before someone else does. Whether it is based on objective reality or not, our fear of scarcity elicits our competitive response, which in turn feeds our fear of scarcity in a self-reinforcing cycle.

The perception of scarcity puts us into a very small mental box. We can expand that box in either of two ways. First, we can realise that quite often the perception of scarcity is not objective but rather of our own making. We can climb out of the mental scarcity box by understanding that there are other seats on the train or bus, and that more buses are coming a few minutes later.

The second way is to decide to step away from the zero-sum paradigm, a rather odd construct when you think about it. Yes, the number of seats on the bus is limited. But another person's gain does not necessarily have to be my loss. Perhaps giving my seat on a bus or train to another allows me to start an unexpected, delightful conversation. Maybe that simple act improves the other person's day or adds joy to mine.

Giving is well known to increase individual happiness, so my 'loss' can actually become my 'gain'. In fact, 'my loss ↔ your gain' can actually become 'our gain'.

It's all about the mindset.

Our mindset is so powerful that it can convince us that a scarcity exists, throwing us into unnecessary competition and thereby objectively creating the scarcity we initially feared. For instance, Tucson, Arizona, is a desert community, and over the years the water has become more and more scarce. The Santa Cruz River, which used to flow freely through the community all year round, is now dry. Only 28 centimetres of rain fall on Tucson each year. And perhaps because water has always been perceived as scarce in this region, the growing population, wanting more, has frantically pumped so much water from the ground that the water table has dropped by more than 91 metres. Trees and other vegetation, which used to line the Santa Cruz, died along with the river itself. The perception of water scarcity, which led to overpumping, then contributed to even greater scarcity, because bare (or paved over) land cannot easily absorb the little rain that falls – most of which is washed away.

Here's the interesting part: the 28 centimetres of rain that Tucson gets each year are actually more than the municipal water it consumes each year.[1] Water was never measurably scarce, it was only perceived as being scarce. Tucson has plenty of water if you consider the abundance of the entire water cycle instead of focusing only on the amount in your well at any given time. When a resource is *perceived* as scarce but is in reality abundant (plenty of seats on a bus or enough rain for everyone), we have the option of reacting either in a narrowly competitive way or in a more broadly collaborative manner. How we react may be influenced by something as profound as our degree of personal self-awareness, or by something as simple as how we happen to be feeling that day. Our attitude does not change any of the facts (how many seats there are on the bus or how much rain falls), but it does make a massive difference in the nature of our experience. And in many cases, when we collaborate, we have more rich experiences, not fewer.

However, when the resources are *actually* scarce and getting scarcer, we face a very different situation in making decisions. Contrary to what we might initially think, in circumstances of real (not only perceived) scarcity, our *only* viable option is collaboration. Fortunately, contrary to what most of us believe, it is the option we human beings tend to adopt, at least under certain circumstances.

In the face of disasters like hurricanes, earthquakes and even terrorist attacks, members of a community tend to come together in solidarity with one another. Studies conducted after Hurricane Katrina in New Orleans and Typhoon Haiyan in the Philippines, as well as many other disasters around the world, have shown that communities respond overwhelmingly with an altruistic spirit of solidarity under the initial common pain and then collaborate to reconstruct and recover afterwards.[2] At these moments, our tendency to give, be it time, skills, money, love or simply a home-cooked meal, overrides our tendency to be competitive. Key to this shift away from competition is that giving makes us happy, so while we act primarily in service to others during times of great hardship, we are also, in fact, acting in service to ourselves.[3]

On 13 November 2015, two weeks before the start of the final session of negotiations for the Paris Agreement, the city suffered its worst terrorist onslaught ever. The attackers targeted six popular locations across the city, killing 130 people and wounding almost 500.[4] No one who was there in the days following will ever forget the sight of thousands of pairs of shoes placed in neat rows in the Place de la République, including a pair of plain black shoes sent by Pope Francis. And far from staying away, 155 heads of state and government travelled to Paris two short weeks afterwards for the largest ever gathering of heads of state and government under one roof on a single day, partly because of the importance of the need to reach a global climate agreement, and partly as a mass demonstration of solidarity with France.

In times of profound suffering and great need, we rise to the occasion, we stand shoulder to shoulder in mutual support. That impulse to gather in a circle of care for one another must be extended to our efforts to address the climate crisis.

Particular recent disasters that you may recall, and the subsequent collaboration and solidarity they precipitated, likely had only a local impact, but the situation we face with global scarcity is vastly more challenging. Globally, we have dramatically fewer insects, birds, and mammals than we did just 50 years ago, and far less forest cover. Our soils are less productive, and our oceans are less bountiful. Harder to see but even more threatening in its consequence is the fact that we are running out of atmospheric space for our greenhouse gas emissions. Think of the world's atmosphere as a bathtub in which, for 50 years, not water but greenhouse gases have been rising. They are now approaching the rim, the limit that the bathtub can hold, or the scientifically established maximum amount of greenhouse gases that the atmosphere can contain – its carbon budget. If we exceed the carbon budget, the bathtub will start to overflow uncontrollably. We are on the verge of atmospheric tipping points that are frighteningly unpredictable and irreversible. Every bit of carbon dioxide (CO_2) emitted – no matter where in the world – contributes to the possibility of disaster. We are rapidly exhausting the space in the bathtub. This is the ultimate scarcity.

Adopted in 1992, the UN Climate Change Convention is based on the recognition that developed countries bear overwhelming historical responsibility for climate change because of the emissions caused by their fossil-fuel-based industrialisation. In contrast, developing countries have insignificant historical responsibility but bear disproportionately high destructive impacts in relation to the size of their economies. That is not ideology, it is an indisputable fact. At the same time, three decades later it is evident that, as they develop and their growing populations emerge from poverty, some developing countries are rapidly increasing their emissions because their economic growth is still largely linked to fossil fuels. As a result, industrialised nations have been advocating that developing countries assume more responsibility for emissions reductions. For years, developing countries have flatly rejected these demands as hindering their economic growth, even as they must shoulder increasing negative impacts from climate change.

Suggestions for a fair allocation of what remains of the carbon budget have been varied. Some have proposed imposing a limit on emissions from industrialised nations so that space remains for those of developing countries; the developed nations deemed this unacceptable. Others have proposed a gradual reduction of emissions in industrialised countries and a managed growth of emissions in developing countries. Unsurprisingly, no happy point of convergence has been agreed on. Another proposal would impose a world-wide limit of two tons of CO_2 emitted per person per year. As the range of national per capita yearly emissions spans from 0.04 to more than 37 tons of CO_2, it was inevitable that those countries substantially above the suggested two tons did not seriously consider the proposal.

Fair allocation of the remaining atmospheric space has proven to be a futile exercise no matter the formula. A fair outcome is not viable as long as we pursue it from a mindset of scarcity and competition.

The state of the planet no longer allows for this mindset because we have reached existential scarcity, limits to the survival of many of the ecosystems that sustain us and that help to maintain safe greenhouse gas levels in the atmosphere.

If the Amazon is destroyed, carbon emissions will rise so high that the entire planet, not only Brazil, will suffer the consequences. Likewise, if the Arctic permafrost thaws, not only will the countries surrounding the North Pole suffer, but so will the whole Earth. We are all in the same boat. A hole at one end of the boat does not mean that only the occupants sitting there will drown. We all win or lose together.

The new zero-sum model presupposes collaboration not competition as the necessary engine for regenerating the biosphere and creating abundance.

It was close to midnight, and we were at our breaking point.

The 2014 negotiations in Lima, Peru, had been moving forwards swiftly over the past days, but now we were at the anticipated impasse: responsibility for emissions reductions. We had known that the issue would raise its head, and that this time the consequences were grave – they would make or break next year's Paris negotiations.

Without fail, at every major international negotiation session, whenever we were on the cusp of an intractable deadlock, there would be a soft knock on the office door, often after midnight, and Minister Xie Zhenhua, for years the head of the Chinese delegation, would walk in. As anticipated, here he was again with a clear message. The draft negotiating text did not properly account for the great differences in responsibility for, and future ability to respond to, climate change. Developing countries would prefer no agreement in Lima or Paris next year, if it meant accepting one that was unfair.

He pointed to a recent agreement between the United States and China that steered away from an approach grounded in competition and scarcity, towards collaboration and abundance. The agreement did not focus on the historical responsibility of industrialised nations nor on the obligations of developing countries to reduce their emissions. It was based on a different paradigm, one that encouraged the shared pursuit of the benefits of emissions reductions for individual nations as well as for the collective: a new model beyond zero-sum.

Now it was our job to adapt that conceptual model to the context of a global agreement between 195 nations in a way that was coherent with all the rest of the issues for which we were finding common ground. First we had to repeatedly negotiate every word and every comma of the adapted text between the US delegation, led by Todd Stern and Sue Biniaz, and the Chinese delegation led by Minister Xie. We had to move quickly but discreetly between delegation offices so as to not give any visible signs of frenzy to the thousands of other delegates who were exhausted and anxious about the deadlock, wondering if the whole session would go up in flames. But after several iterations of goodwill on both parts, an agreed version emerged, and each side undertook to bring their respective group of countries on board.

The new understanding established that reducing emissions is indeed a responsibility of every nation, for its own enlightened self-interest *as well as* for the benefit of the planet as a whole. The mindset shift and associated new language in the text – away from competition and towards shared winning, where everyone can gain from a new abundance without impinging on each other – unlocked the door to the global agreement that would be signed in Paris the following year.

An increasing number of countries today fully understand that their development in the twenty-first century can and should be clean; that by decarbonising their economies, they can reap the benefits of more jobs, cleaner air, more efficient transportation, more habitable cities and more fertile lands. This shift towards a mindset of creating abundance does not negate the limitations of a carbon economy; instead, it gives every country a wealth of positive individual and collective reasons to stay within that limit. As one country moves forwards demonstrating the national benefits of clean technologies and policies, others will follow, momentum will be built, and the global rate of decarbonisation will increase, protecting the planet.

When we are motivated by a desire for collaboration, we liberate ourselves from the restrictive framing of attaining 'what I want, or think I need', and open ourselves up to a broader framing of what is available and possible in many other forms – available to me, but not only to me, to others as well. The realisation of abundance is not an illusory increase in physical resources, but rather an awareness of a broad array of ways to satisfy needs and wants so that everyone is content. In this way resources will be protected and replenished, and the relationships among us are enriched.

Endless abundance.

At the individual level, we are called to enhance collaboration and nurture abundance as a mindset. Making that mindset shift is not as hard as it sounds. Consider, for example, the endless abundance of energy coming from the sun, wind, water, sea waves and heat within the Earth, all of which we are now harnessing to produce electricity, and none of which will ever get used up. Regenerated soils, forests, and oceans can all be wisely managed for endless abundance rather than squandered for imminent depletion. In fact, ecosystems operate from the very principle of abundance – they depend on components within them that are naturally plentiful, such as waste, to provide the food and nutrients for further growth.

We can also add creativity, solidarity, innovation and many other abundant human attributes available to us, endlessly.

The rise of collectively generated and freely shared knowledge on the internet has data challenges that remain to be addressed, but it has made the notion of collaborative systems and endless abundance easier to understand. Think of Wikipedia, LinkedIn or Waze. Each user of the system is unique, but all users are interrelated through the network of the endlessly growing system. Every user contributes to the whole, but the total body of knowledge is larger than the sum of all users. And the system is in constant change, amplifying in some areas, correcting course in others and growing into previously unknown spaces. Competition plays a role, but it is limited because everyone contributes, everyone benefits and everyone partakes of a constantly increasing total. Collaboration is the name of the game. Shared benefit from endless abundance is the result of the game.

As a next step, one could imagine a world of 'open source everything', an open approach in every field of human endeavour, where competition is no longer the operating principle, but rather collaboration. Following the principles we observe in any natural ecosystem, this approach explicitly promotes learning and growth throughout the whole system. It allows us to constantly teach one another, thereby exponentially increasing our capacity to co-create knowledge and share goods and services with open access, used by everyone for the benefit of all.

The practice of abundance starts by shifting our minds away from perceived scarcity to what we can collectively *make* abundant. In so doing, we will become more aware of others, what we can learn from them and share with them. We will be more conscious of our own impulse to compete and, as a corrective, develop a keener interest in how we can all win. We will be more likely to show appreciation to those who have contributed to a joint task, encouraging ever-higher levels of teamwork and collaboration everywhere. We will share the results of our labour with anyone who can use them as input to their further work, without mentally claiming any intellectual property rights. Another person's success is not our loss; it is our constantly growing collective success.

We are entering the next phase of human evolution. The human species (and many other animal and plant species) must now adapt to the scarcity of natural resources we have caused, and the rapidly diminishing space left in our global atmosphere for carbon emissions. To do this, we need to prioritise collaboration. Faced with the ultimate scarcity, we must internalise the new zero-sum (either we all win or we all lose) and apply a mindset of abundance to that which we have left and that which we can co-create and share.

CHAPTER SEVEN

Radical Regeneration

Exhausted after a long day's work at the UN, we were having a quiet meal at a little restaurant close to our office, chatting and commenting on what had been done and what was left to do. Two young men sitting next to us had finished eating and were talking over their third beer about what to do next. We tried to focus on our own to-do list, but their conversation pulled us away.

'But why do you want to leave?'

'Because there's nothing more for me here.'

'So where do you want to go?'

'I don't know. Wherever I can get something better.'

We looked at each other with raised eyebrows. The man had expressed a sentiment we'd heard so many times before – that when there's nothing left, it's time to find more elsewhere.

The man's focus on 'getting something better' was no individual quirk. It has been with human societies for centuries. Conquerors of distant lands pillaged colonies for metals, minerals, and exotic foods, in many cases leaving little more than chaos, infectious diseases, and Bibles in exchange. As managers of fertile soils, we humans have proved remarkably effective at extracting trees and nutrients, leaving only depleted topsoil in our wake.

There's nothing inherently wrong with these instincts. They help us grow to meet rising challenges. But our growth, both personal and professional, is a two-way street: what we get *and* what we give. As a species, however, we have become used to a one-way transaction, that of *getting*, often losing sight of the void that our taking has created.

Our planet can no longer support one-directional growth. We have come to the end of humanity's extraction road. The time for 'getting' is over. Staring us in the face is a huge red sign that reads STOP: PRECIPICE AHEAD.

Extraction is a propensity deeply ingrained in human behaviour. To move away from extracting and depleting, we need to concentrate on another equally strong and intrinsic trait: our capacity for supporting regeneration. Caring for ourselves and others. Connecting with nature. Working together to replenish what we use and to make sure plenty remains for tomorrow. These tendencies are just as much second nature, but they are less well-developed in modern society. It's time to bring them to the surface.

Being regenerative is not strange to us.

If you have children, think about how supportive you are with them when they go through periods of deep doubt. Remember how patiently you listen to their worries and help them stay hopeful. Or think of how encouraging you are to friends who may have fallen into a professional hole, how much time and energy you invest in helping them replenish their self-confidence so that they can rise to the top of their game once again.

Sometimes it's easier to act in more regenerative ways with our friends and families – or even with strangers halfway across the world – than with ourselves. While this may be noble, to be most effective, we need to begin with ourselves.

Amid the climate crisis, we each have an urgent responsibility to replenish ourselves and protect ourselves from breaking down. In the face of imminent burnout, some of our colleagues who have worked for years to address climate change under extremely stressful circumstances have at some point prudently taken time off to restore their energies by turning to the healing arms of nature or the loving embrace of a spiritual community. The wisest among them have incorporated meditation and mindfulness practices into their daily lives.

We know from our own experience that continual personal grounding is key to being able to withstand the daily bombardment of bad news from all sides. Without such grounding, you will be a leaf in the wind – vulnerable to the elements from all directions. Better to stand as a tree, firmly rooted in your own values, principles and convictions. The two of us easily notice the difference between a day in which we meditate and a day in which we don't. The benefits of meditation undoubtedly blossom with years of practice, but they are also palpable on a day-to-day basis. Maybe you don't care for meditation, and a spiritual practice holds no interest for you. Fair enough. But this does not mean you should not be mindful of yourself. Whether it is gardening, crafting, drawing, playing or listening to music, exercising, meandering in the park or paddling down a river, identify what replenishes you and your soul, and do it regularly and intentionally.

Our first responsibility is to notice how and when we are depleted and to support ourselves. Our second responsibility is to reaffirm and strengthen the regenerative capacity we already display with family and friends. But we cannot stop there. Our third responsibility is to engage those beyond our innermost circle and, indeed, nature itself.

In the natural world, the strictest interpretation of the term *regeneration* is the self-generated healing process that restores an organism's injured bodily part from the remaining healthy tissue. For instance, newts, lizards, octopuses and starfish have the capacity to regenerate lost limbs or tails. In humans, adults can regenerate a damaged liver to its original size after either partial removal or injury. And all of us have witnessed the miracle of skin repairing itself after a scrape or wound, sometimes leaving no trace of the injury at all.

A broader interpretation of regeneration is the capacity of a species or a biosystem to recover on its own, once humans remove the pressure they had been exerting. Whale populations and degraded lands are good examples. Grey whales and humpbacks, once decimated by nineteenth-century commercial whaling practices, have now almost recuperated their numbers. The prohibition of whaling shows that if we remove the extractive pressure, animal populations have the ability to rebound (assuming of course we have not driven them to extinction). The same is true for ecosystems, as we can see in photos of ancient ruins abandoned by humans that have been taken over by the surrounding green growth. The recuperation of a flourishing ecosystem around Chernobyl is a great example. With humans gone, the plants started to grow back, supporting worms and fungi that nourished the soil. Birdsong is now abundant and even large mammals like boars and bears have returned. If we remove the pressures we have wielded, nature tends to return to health.

The converging crises of climate change, deforestation, biodiversity loss, desertification and acidification of the oceans have taken us to the point where we can no longer naïvely depend on the Earth's natural resilience or capacity to recuperate. While nature is innately restorative, regeneration does not always occur completely on its own. Right now, we have almost extinguished nature's capacity for self-renewal. In many cases, ecosystem restoration requires intentional human intervention, such as *rewilding*, by which we not only remove the destructive pressure of grazing or unsustainable harvesting but also reintroduce native animals and help nature bounce back, slowly recuperating its rich biodiversity. Planting trees and shrubs in degraded or deforested landscapes is an intentional regenerative process that restores soil health, increases productivity and stabilises underground aquifers. In one well-known effort currently under way to reforest the Scottish Highlands, researchers noticed that when the trees were lost from the landscape, so were the fungi normally found in the soil around them. It turns out that mycorrhizal fungi are hugely beneficial for reforesting degraded landscapes, and now a sprinkling of native mushroom spores is added to the roots of new saplings as they are planted to speed up and strengthen the revival of the Great Caledonian Forest.

Coral farming, another fine example of intentional regeneration, is the process whereby fragments of corals are collected from local reefs, further broken up, raised in nurseries where they mature much faster than in the open sea, and then planted at the restoration site to regrow the damaged reef. With the advent of innovative coral farming techniques, scientists will soon be able to launch large-scale restoration efforts to revive the valuable coral reefs that are at risk or already dead. Nature can restore itself, but with intentional human help it has a better chance and can speed up. With our support, regeneration can become the predominant direction of the future evolution of this planet.

We have brought our natural world to several perilous brinks from which it may not be able to recover on its own. It is like an elastic band that stretches and contracts normally but if stretched too far will snap. Undoubtedly the regeneration of nature now needs to be intentional, planned and well-executed at scale.

We will not recover everything. Many species are already extinct and will not return, and some ecosystems may already be damaged beyond their resilience threshold. But fortunately we still have a relatively hardy natural environment that responds positively to our care and caring. Well-intentioned and well-planned regenerative practices will restore our ecosystems, perhaps not to their former state but to a new state of regained health with enhanced resilience.

Let's begin our regenerative mindset shift by acknowledging and internalising the simple fact that our lives, our very physical survival, depend directly on nature. Human beings cannot survive longer than a few minutes without oxygen. The oxygen we breathe comes from the photosynthetic processes of trees, grasses and other plants on land and of phytoplankton in the oceans. Every sip of water we drink comes from rain, glaciers, lakes and rivers. Without land we would have no food to eat, no fruits, vegetables or grains, no cows, chickens or sheep; and without rivers and oceans, we would have no fish or seafood to consume. Humans cannot survive for more than a week without water or for three weeks without food. Every breath we take, every drop of liquid we drink, and every morsel of food we eat comes from nature and connects us profoundly to it. It is a simple basic truth, yet one we often tend to ignore or take for granted.

It is not only our immediate survival that depends on functioning ecosystems. In large part our health, physical and emotional, relies on having contact with the natural world around us. This contact is under threat from rising rates of urbanisation and from time spent with our electronic devices. Sedentary indoor life – often characterised by limited natural light, poor air quality, walled surroundings and increasing screen time – leads not only to obesity and loss of physical strength but also to feelings of isolation and depression. This family of symptoms has been broadly diagnosed as 'nature deficit disorder'.[1] Conversely, studies show a significant decrease in mortality, stress and illness for those who exercise and spend time in the natural world. Nature-based play, gardening and access to natural landscapes heighten our sense of well-being while sensitising us to the ever-changing light, weather and seasons.

Reconnection to nature is a powerful antidote to anxiety and stress, as well as a counter to physical illnesses. The Japanese health system has developed the practice of *shinrin-yoku* – literally, forest 'bath' (not with water) – or spending mindful time in the woods. It is beneficial for soul and body as it boosts the immune system, lowers blood pressure, aids sleep, improves mood and increases personal energy. It has become a cornerstone of preventive health care and healing in Japan.

A growing number of paediatricians are prescribing more unstructured time in nature for children to fight childhood obesity while engendering a sense of wonder and love of local wildlife, fauna and special places. In fact, some doctors argue that watching documentaries about endangered species and faraway ecosystems cannot substitute for personally caring for plants at home and directly exploring the flights of butterflies, birds and dragonflies.

Public consciousness of our dependence on, and interconnectedness with, the planetary life support system is growing, along with an increasing awareness of the need to restore ecosystems and planetary health. Countless efforts are underway around the world to plant trees, protect mangroves and peatlands, re-establish wetlands, and restore degraded lands via rain harvesting, perennial grains, grasses, and agroforestry. But more is needed so that these solutions can be taken to scale globally.

A regenerative mindset is most effective if pursued intentionally and consistently. It is both a tough mental discipline and a gentleness of spirit that needs to be cultivated. It is about understanding that beyond getting what we want and need from our fellow human beings, we have the responsibility to replenish ourselves and to help others to restore themselves to levels of greater energy and insight. It is about understanding that beyond extracting and harvesting what we need from nature, it is our responsibility and in our enlightened self-interest to protect life on this planet, indeed even enhance the planet's life-giving capacity. Personal and environmental goals are interlinked, mutually reinforcing, and they both need our attention.

A regenerative mindset bridges the gap between how nature works (regeneration) and how we humans have organised our lives (extraction).[2] It allows us to 'redesign human presence on Earth'[3] driven by human creativity, problem solving and a fierce love of this planet.

Sir David Attenborough, one of the most renowned naturalists of our time, has warned us that 'the Garden of Eden is no more'. We agree. That is why we now have to create a Garden of Intention – a deliberately regenerative Anthropocene.

Instead of strip-mined mountains, destroyed forests and depleted oceans, imagine millions of rewilding projects covering over a billion hectares of forests, regenerating wetlands and grasslands, and restoring coral farms in all tropical oceans.

We will not have a regenerative Anthropocene by default, but we can create it by design. With directional intent, we can shift our aspirations from our current extractive growth to a life-sustaining society of regenerative values, principles and practices.

We can ignite regenerative human cultures that seek to ensure that humanity becomes a life-sustaining influence on all ecosystems and on the planet as a whole. We will need artists as well as policy experts, farmers as well as leaders of industry, grandmothers as well as inventors, and indigenous leaders as well as scientists.

We can choose regeneration as the overarching design principle of our lives and our activities. We can restore the resilience of the land and our communities while healing our souls. Our corporate strategy meetings and family reunions should be carbon neutral for sure, but beyond that, they can include regenerative projects in which we put our hands in the soil or in the water, together taking actions that restore rather than degrade life on our planet.

We have to shift our action compass from self-centric to nature-aligned. We have to filter every action through a consequential stress test, and we have to be pretty radical about it. When considering an action, we have to ask: does it actively contribute to humans and nature thriving together as one integrated system on this planet? If yes, green light. If not, red light. Period.

This is not a distant dream. It is already happening. Together with renowned author Arundhati Roy, we can say, 'Another world is not only possible, she is on her way. Maybe many of us won't be here to greet her, but on a quiet day, if I listen very carefully, I can hear her breathing.'

PART III
TEN ACTIONS

CHAPTER EIGHT

Doing What Is Necessary

Towards the end of the first week of the Paris negotiations in December 2015, we were working in Christiana's office when we heard a knock on the door.

Kevin O Hanlon, head of UN Security, came in. We had all worked together for years, so the concern on his face was easy to read.

'We found a bomb.'

It was the nightmare scenario we had been dreading.

Because of the recent terrorist attacks in Paris, we had allowed the security forces of the host country to assume responsibility for the arrival and departure area of the UN meeting grounds. By law, the location of a UN negotiation meeting is considered extraterritorial for the duration of the meeting, therefore not under the sovereignty of the host country. But for COP21, we had transformed Le Bourget Airport into a large conference centre, and with 195 countries and 25,000 people in attendance, it was an obvious potential target. We knew we needed help from French law enforcement, especially the specialised French anti-terrorism police and their bomb-sniffing dogs.

Thirty thousand police officers had been deployed across the country, and 238 security checkpoints had been set up. Security was unprecedented. What we were attempting to accomplish inside the UN grounds was unprecedented as well. Now we were five days into the largest climate change negotiations in UN history. The stakes were enormous.

Kevin explained that the bomb had been found in a trash bag in the transportation hub of the Le Bourget subway station, the main train stop to our conference – every single one of the 25,000 participants streamed through that station all day long. Christiana's two daughters used the station at least twice a day. Tom had two children at home, waiting for him to return. We looked at each other and saw in each other's eyes the scenes from three weeks earlier in Paris and Saint-Denis. Broken glass. Blood. Dead bodies. Family members weeping.

The bomb had been deactivated, but there was no way to determine if there were more explosive devices in the area.

Everything hung in the balance. After years of development, we finally had a draft text of a global climate agreement. We had the long-term target of a net-zero emissions economy, language to protect the vulnerable and even a ratchet mechanism to periodically deepen emissions reductions to try to keep the world to 'well below 2 degrees Celsius' of temperature rise. These ambitious goals were in the draft text but were not guaranteed to survive many countries' political pressure to remove them. Plus, we wanted more. We wanted the agreement to put us on a path to a 1.5-degree-Celsius maximum temperature rise. A 2-degree world would result in up to three times as much infrastructure destruction, biological destruction, and life-threatening heat, hunger and water scarcity. The difference would save millions of lives and perhaps even give low-lying islands and coastlines a chance of survival. If we called off the conference, we didn't know whether we could ever achieve an agreement again – formidable political obstacles remained, and the forces of resistance were beginning to gather to prevent the world from achieving what it needed to do.

This was our chance.

And now a decision was needed.

Should we close down the conference and with it the chance for a global climate agreement, or should we keep it open, with all the risk that this entailed? Christiana was no stranger to making hard choices, but this wasn't a choice a mother should ever have to make.

All the risks, the fears and the loss washed over us both in that moment. It was a terrifying place to be, but we couldn't stay there long. We had to act – one way or another.

You also have a choice ahead of you, and by now you understand the risks.

The time you have to make that choice and act on it is vanishingly small. We have discussed the mindset everyone needs to cultivate in order to meet the global challenge of the climate crisis, but on its own, this is not enough. For change to become transformational, our change in mindset must manifest in our actions.

There are ten necessary actions for the making of a regenerative future, the future we hope you will choose. Some may be familiar; others will be new. We have considered not only the world we are trying to create but also the risks inherent in the effort.

On one level, the big solution to the climate crisis is blindingly obvious; we need to stop filling our atmosphere with greenhouse gases. But in order to deliver on that goal, we need to find myriad small solutions.

Greenhouse gases are emitted as a direct result of the things humans do to survive, such as sourcing food and getting around. Our ways of doing and being have become so entangled with what is killing the planet that we cannot feasibly just flip a switch and stop emitting greenhouse gases.[1] Consider the implications: if in an imaginary world, we stopped using all fossil fuels in an instant, if we denied people what they are used to – we would have a global revolution in a matter of weeks if not days.

On the other hand, if governments do not do enough and keep endangering the lives of young people and their future children, a major uprising is also likely and perhaps even already underway.[2]

We need transformational change at the speed that science demands and in a manner consistent with democracy – that is, if we do not wish to descend into tyranny or anarchy.

This point is critical. In the coming decades, climate change will show up in larger and more lethal ways, leading to more forced migrations, changes in agricultural output and more extreme weather. Increasingly populist leaders will try to justify their actions by claiming to protect the short-term interests of those they govern. This could hinder attempts to deal with the root causes of climate change, thereby worsening the crisis. Even the most casual observers of today's politics see that this risk is not merely theoretical. A five-year drought in Syria – the worst ever recorded – destroyed agriculture and caused many rural families to migrate to cities. Large numbers of refugees were already pouring in from the war in Iraq, and the combined tensions gave rise to the civil war and the atrocities committed by Bashar al-Assad. Then a flow of refugees, largely from Syria, made their way to Europe, where Chancellor Angela Merkel eventually accepted many into Germany.[3] This led to fundamental changes in the German political system as the AfD (Alternative for Germany), a far-right movement, jumped from averaging 3 per cent in the polls to 16 per cent and is now a major political force.[4] This weakened Merkel, then the de facto leader of the European Union, and it continues to affect politics in Europe and beyond.

If we are to resist extremist politics as the effects of climate change grow ever more critical, we will have to be vastly better prepared than we are today. The ten action areas we set out here attempt to portray not only how we can reduce emissions but also how as a society we can make ourselves more resilient to extremist movements that could pull us back in the wrong direction.

The ten actions that we call for are not only about moving beyond fossil fuels and investing in technological solutions. They also call for a fairer economic system that does not strain the social net even further. They call for strong political engagement by everyone, and for relinquishing nostalgia for a past that might be dangerous to re-create. The additional pieces may feel remote from the issue of climate change, but they are fundamental parts of our response. We must reject the cycle of blame and retribution, and embrace the shared endeavour we so desperately need. We cannot strain the social safety net and continue to expand inequality, or else our democratic systems will refuse to allow further changes to the economy. We have to get our arms around the whole issue at the same time.

What we will ask of you is significant. It is not simply about making minor changes to lifestyle, although those can be important too; it is about transforming our priorities in order to create a future in which all of us may thrive. It will involve developing and utilising the qualities of mind we talked about in the previous section and using them to take greater steps towards creating a new world.

None of us has complete control over which path the world ultimately chooses to take and which future will be ours. But each of us individually can engage in these ten action areas, giving direction to the transformation towards a regenerative world.

We are all weavers of the grand tapestry of history. As we cast our minds back and consider those who lived at moments of great consequence, we naturally feel that if we had lived then, we would have been among those who made the noble choices rather than those who stumbled along, head down, changing nothing. Well, this is our chance. Every one of the needed actions is something you can personally achieve as a human being, even if that boils down to urging others to take it seriously. Our hope is that by the time you put this book down, you will understand that you can make a significant difference.

We can no longer afford the indulgence of feeling powerless.

We can no longer afford to assume that addressing climate change is the sole responsibility of national or local governments, or corporations or individuals. This is an everyone-everywhere mission in which we all must individually and collectively assume responsibility. You play many roles in your life – parent, spouse, friend, professional, person of faith, agnostic. You may have great means or none at all. You may sit on the board of a corporation or lead a city, province, or country. Whoever you are, you are needed now in every one of your roles.

Changing our mindset is critical but does not suffice. We invite you to dive into *doing* as soon as possible. Focus on doing one or two of the ten actions at first. Choose the areas that make the most sense for you, and then challenge yourself to do more over time. Know that our discussion can only point the way, shining a light on what we think is critical at this unique moment, but all of us can do myriad other things to make a difference.[5] If you leave this book with a commitment to be part of this journey, then you will need to go beyond what we set out here.

You already know the end of our bomb story. We had to do what was necessary, no matter the cost. We knew the only way to truly protect our own children was to courageously continue the work of protecting all humanity and our planetary home. The metro station stayed open. The conference proceeded. Taking this action was not without risk, but neither of us regrets it. Hopefully, in ten years, we will be able to say the same about our collective action.

The time for doing what we can has passed.

Each of us must now do what is necessary.

Action 1
Let Go of the Old World

To meet the challenges of the climate crisis and preserve all that we hold dear; to retain democracy, social justice, human rights, and other hard-won freedoms in the future, we must part ways with that which threatens to destroy them. Now is the time to make profound shifts in how we live, work and relate to each other. To be successful, we need to make a series of intentional moves.

The first of these is to honour the past, then let it go.

Fossil fuels have given a huge boost to humanity's development, but their continued use is no longer supportable because of the extraordinary damage they cause to our health, our ecosystems and our climate. Viable alternatives are safer. Now is the time for us to thank fossil fuels, retire them and move on.

It is the same story for so many of the profound shifts we need to make today. The building blocks of our current society – energy, transportation and agricultural systems, which we now know to be harmful – must undergo radical transformations.

We all find change difficult. We tend to cling to what we know and resist what is new – even when the new brings tremendous benefits. Opposition to onshore wind turbines in the UK is a good example. Even though onshore wind is now the cheapest form of energy[6] (cheaper than coal, oil, gas, and other renewable sources), rural landowners have significantly resisted it, keen to preserve the appearance of the countryside. When the Conservative Party (which derives much of its support from these rural communities) came to power in 2015, it slashed subsidies and changed planning laws for onshore wind – leading to an 80 per cent reduction in new capacity.[7] Only now, with climate change awareness rapidly rising among the UK public, is support for onshore wind starting to outweigh an attachment to yesterday's aesthetics.

Be mindful that some individuals and industries are actively fighting the changes we need to achieve a world that is only 1.5 degrees Celsius warmer. They are sowing fear and uncertainty, sponsoring divisiveness, and seducing us into an unconstructive blame game, all of which we would do well to resist.

Change makes us vulnerable to tribalism and to the illusion of certainty. In the transition to a regenerative world, one of the biggest risks is that the political centre does not hold, and people succumb to the easy promises of populist leaders at either end of the political spectrum. History and early signs both suggest that this might be our new reality with the real potential to turn democracy into tyranny. We cannot go back to the way of life that created the climate emergency in the first place, but treading new ground is politically challenging. The political shocks currently reverberating across the world are just the start.

Change can also trigger blame. Some people who claim to be on the right side of the climate change debate will have a narrative laced with exclusion or blame. Blame is already a powerful current in our relationship to climate change – it is directed towards the developed world, the oil industry, capitalism and corporations, particular countries, and the older generation. Outrage is understandable, particularly now that we know beyond doubt that some companies hid the truth about climate change for decades in order to continue making money.[8] In those cases, justice and due process are called for and should certainly be delivered.

But blame does not serve us. It creates a sense of needed restitution but does not actually deliver it. Blame can consume us and cause us to lose years of constructive action. History shows very clearly that once humans start pointing the finger of blame at each other, it can be hard to stop. In the aftermath of the First World War, the Allied powers humiliated Germany, forced her to accept full blame for the war, and imposed crippling reparations payments. Historians agree that that paved the way for the rise of fascism and a second massive global conflict 20 years later.[9]

Here's what we can do to let go of the old world and keep the worst of our impulses in check:

Focus on where you're going, not on where you've been. Cultivate your constructive vision for the future and hold on to it, come what may. When you can see where you're going, you won't be so afraid of losing your grip on the past.

Build resilience to nostalgia. Recognise and understand the inherent impermanence of our world, and build a practice of non-attachment. We can all be susceptible to a desire to re-create the past. However, history teaches us that at moments of profound change, our nostalgia can be used against us. It can distract us from the urgent work ahead, and political leaders may appeal to the past to manipulate our emotions and secure our consent to act immorally.

Burst out of your bubble. We will not be able to make big changes in our society without fully understanding and accepting one another's deeply held values and legitimate concerns. Certain segments of our society may continue to resist change for good reasons, and our failure to understand them may set us all back. In 2018, French President Emmanuel Macron tried to approach reducing emissions and air pollution by increasing the fuel tax. But he failed to bring everyone on board – those struggling to make ends meet faced unacceptable increases in the cost of their commutes. The result was a fury of protest, catching the government completely off guard. And the French *gilets jaunes* (yellow jackets) activists spectacularly forced Macron to abandon his plan.[10] Why do these disconnects happen? Partly because we are becoming increasingly divided by the type of media we consume. We tend to read opinion pieces that reflect or support our own views, reinforcing what we want to hear and already believe. Cleverly programmed algorithms turbocharge that process on the internet and social media.[11]

This means that often we have no idea what other people deeply value or think.

Get offline and get to know your neighbours, people in the supermarket queue or fellow commuters. Challenge your own assumptions, and be mindful of misinformation and disinformation. Share your hopes and fears *in person,* listen to others, and be honest and respectful.

In 1990, after spending 27 years in prison, Nelson Mandela was informed by President F. W. de Klerk that he would be freed in less than 24 hours. The following day Mandela walked out of Victor Verster Prison and into history. He had to pass through a courtyard, beyond which he would be a free man. As he later recounted, he knew that if he did not forgive his captors before he reached the outer wall, he never would. So he forgave them. This did not mean that he forgot. The Truth and Reconciliation Commission (TRC) that he later established played a remarkable role in helping post-apartheid South Africa let go of its past. The TRC allowed anyone who had been a victim of violence to be heard in a formal setting. In addition, anyone who had perpetrated violence could also give testimony and request amnesty from prosecution. Mandela's achievement and the process he established greatly aided the transition from one state to another very different one.

The past was relinquished, and the future finally had room to emerge.

We too must let go of the fossil-fuel-dominated past without recrimination. The process of letting go is essential, and it must be intentional. The more work we do to let go of the old world and walk with confidence into the future, the stronger we'll be for what lies ahead.

Action 2

Face Your Grief but Hold a
Vision of the Future

The winters, springs, summers, and autumns, the rainy and dry seasons that we remember will not be those that our children and their children will enjoy. It's rare today to find someone over 50 who isn't conscious that the weather patterns that defined their childhoods are being quickly and drastically altered. Glaciers and lakes are rapidly retreating, and our oceans are choking in plastic.[12] Ancient bones and diseases are surfacing in the permafrost.[13] As our weather and landscapes change before our eyes, as millennial signposts of natural rhythms disappear, our understanding of the ways of the world is unravelling. Things don't make sense the way they used to.

We cannot hide from the grief that flows from the loss of biodiversity and impoverished lives of future generations. We have to feel the full force of this new reality in our bones. There is a power to consciously bearing witness to all that is unfolding without turning away, and counterintuitively, you may actually feel better about the situation when you deeply accept the reality of it. And beyond this, we also then need to look to the future and set our sights on what we can still create. The changes to come will be more disorienting than those we have already experienced, and it will be easy to lose our footing unless we can clearly see where we want to go.

We need to take responsibility for this reality by facing the uncertain future with as much courage as we can muster. Doing so requires us to understand why we must meet this moment with energy and commitment.

For years, the countries of the world tried to reach a global agreement on climate change. The effort became so all-encompassing that the challenge being attempted began to merge with the reason for doing it. The vision became securing a global agreement. As powerful and important as it was, the global agreement was actually a goal in service of a vision. The vision was, and still is, a regenerative world where humans and nature can thrive.

Confusing vision with goals is easy. A goal is a specific target that we set on the way to achieving a vision. It includes the strategies and tactics we use in moving towards the vision. Goals are critical, but we also need a vision to inspire the kind of commitment and energy we will need to get through the difficult years ahead. If we don't have a vision, our goals alone may not afford us the flexibility necessary to achieve the vision.

And if we lose sight of the big picture and become fixated on how to achieve it, progress can grind to a halt or, worse, divisiveness can take hold.

However, for those eager to take action, fixating on the vision can feel irresponsible and unconnected to reality. When we are caught up in the issues of today – communities decimated by increasingly violent weather patterns; the unbridgeable chasm between the rich and the poor; rapacious multinational companies focused on short-term profits rather than long-term value; and political leaders bent on driving divisions between nations (and within nations) – having a vision can seem naïve and wishful thinking. The distance between projecting a vision of a better world and realising it through concerted action can sometimes seem unbridgeable.

Having a vision is essential, but we have to be open to doing things in new ways. So hold on to your vision, but remain flexible and adaptive about the route to get there. The route may change based on circumstances, while the vision remains a fixed North Star, a guide *and* a destination.

Start with why. You do not have to believe your vision is likely to be achieved, or that the struggle to achieve it is going well, to keep pursuing it.

Pondering the different scenarios presented at the beginning of this book, you may conclude that we cannot turn this ship around in time, that we are going to crash, and that our vision is unattainable. That thought is not irrational. What would be irrational is to imagine that the reasons for building a better future are therefore diminished. Stubborn optimism needs to motivate you daily; you always need to bear in mind why you feel the future is worth fighting for. The essential 'why' should be the driving force of all efforts to combat climate change no matter what.

Imagination is essential. Ideologies and ways of organising this world can seem very ingrained, but they are subject to major disruption more easily than you think. It took Emmeline Pankhurst and the suffragette movement slightly more than a decade to force the British government to allow women the right to vote.[14] The Soviet Union seemed so solid as to be eternal, but once cracks started to appear, the edifice crumbled in just a few months.[15]

In 1939, General Motors presented visitors to the World's Fair in New York City with an imaginative vision of what the future could look like. It was called Futurama and consisted of an enormous model of multiple high-rise buildings, vast suburbs, and large motorways for travel between them, necessitating the use of cars.[16]

Imagination is going to be critical as we work to transform today's urban sprawl to make it fit for the future. Some futurists have predicted that in the course of a decade, the rise of the autonomous, shared, on-demand electric car means we will need 80 per cent fewer cars on the roads than we do now.[17] This will free up huge areas of urban space that are currently used as parking lots.

In London, for instance, it could mean that 70 per cent of the space currently used for parking cars, or the equivalent of about 5,000 sports fields, could become available for growing food, rewilding or building sustainable housing.[18]

Much of what we imagine to be permanent is more ephemeral than we realise. Sometimes imagination can seem naïve, but don't belittle thinking big. Time and again societies have turned seeming fantasies into realities when circumstances require something new.

Keep your eyes on what's to come. There will be times when we feel we are failing. However much we progress, we will see some deterioration in our environment and our society. Heartbreakingly, people will die as a result of climate change, land that people live on will become uninhabitable, and species will continue to become extinct – all causes for real grief, and grieving is needed. Give adequate time and space for that necessary mourning, and seek support from your communities – both are extremely important. We cannot and should not turn away from the pain, but that heartbreak should spur us onto greater action rather than sink us into a pit of blame, despair or hopelessness.

As Maya Angelou said so eloquently: 'You may encounter many defeats, but you must not be defeated. In fact, it may be necessary to encounter the defeats, so you can know who you are, what you can rise from, how you can still come out of it.'[19]

A compelling vision is like a hook in the future. It connects you to the pockets of possibility that are emerging and helps you pull them into the present. Hold on to that. Stay firmly fixed to a vision of a world you know is possible. This act is radical resistance to the belief that solving our problems is beyond us.

When Martin Luther King, Jr., stood on the steps of the Lincoln Memorial in August 1963, the outlook for race relations in the United States was grim. Just months earlier, Alabama governor George Wallace had stood outside the Alabama state capitol and declared 'Segregation now, segregation tomorrow, segregation forever'. To enforce segregation, police unleashed dogs and water cannons on protesters, even on children as young as six. Even those who supported civil rights felt that change was too far off and the campaign was hopeless. Given that context, King's words about having a dream were like a light in darkness. He didn't know how it was going to happen, but he held tight to his vision of a society in which people were treated equally regardless of their race. The following year his persistence led to the passage of the Civil Rights Act, and his vision lived on after his death, inspiring equal rights movements across the world and embedding non-violent protest as a cornerstone of political protest movements.[20]

A world that has become richer in the active use of vision and imagination is a much more vibrant, inspiring and joyful place. In these complex times, we often lament the lack of global leaders who can show us the way and help guide us. Those people are important, but we must all believe that the world *is* worth saving and a regenerative future is utterly possible. In the end, we are not going to solve this problem by hoping that our democratic systems produce enlightened leadership. They might, but the survival of our species can't depend on the partisan lines of a divided electorate. Instead, we must all embrace a strong vision of a better future.

Action 3
Defend the Truth

Three centuries ago Jonathan Swift wrote, 'Falsehood flies, and truth comes limping after it.'[21] How prophetic this turned out to be. A recent analysis by MIT shows that on Twitter lies spread on average six times faster than truth, and that truth never reaches the same level of penetration.[22] Social media is an engine for the production and dissemination of lies.

This fact has serious consequences for our society and in particular for our ability to come together to deal with complicated long-term threats like the climate crisis. In this 'post-truth era', the undermining of science now has currency.

The fabric of the scientific method is fraying. Objectivity is under attack. Some political leaders have chosen to part company with objective reality. The rise of social media has afforded these leaders ample opportunity to obscure facts. This move towards subjectivity creates a breeding ground for oppression and tyranny. We all have an urgent responsibility to recognise and defend such an attack on truth because if it persists, our small window of opportunity to turn back the tide on the climate crisis will be lost forever.

In no period of history did leaders ever speak the truth at all times, but right now an altogether different level of lying is evident in the political arena.

Humans are vulnerable to the post-truth world for a reason. Our natural inclination seems to be to seek confirmation of things we already believe to be true, rather than evidence for an objective reality.[23]

It feels good to have our beliefs confirmed, and we respond with positive emotion to anyone who makes us feel this way. Thus, if a leader affirms our belief that vaccines cause autism, or that climate change is a hoax, or that anything else that we feel to be true is true, then we get a frisson of positive emotion. This well-documented and -researched phenomenon is called confirmation bias.[24]

Climate change will result in disasters, lots of them: inundations of major cities, loss of islands, a rising tide of migration. At these moments of extreme vulnerability, leaders with authoritarian instincts will want to seize the chance to consolidate their power. Populist authoritarian rulers will not seek to address the complex climate crisis with long-term solutions; instead they will find someone to blame. We cannot allow them to use the coming disasters to exacerbate tragedy to the detriment of us all.

Here's what we can do to defend the truth:

Free your mind. In the end, you are responsible for what you choose to believe in a post-truth world. Make no mistake, this problem is not ancillary to the climate crisis. If we can't agree on something as basic as a verified fact, our hands will be tied when it comes to the big stuff, and climate change is *huge*.

The reality of climate change is finally provoking genuine public anger, spurring people onto the streets. Our democratic systems cannot resist our voices for long, provided we can maintain the basis of objective truth within our societies. We must consciously enter into a state of self-reflection, questioning whether we are making a conscious choice to adhere only to information that does not challenge our position. For example, the fact that you are reading this book might be an instance of your own confirmation bias. Pay attention to your own eagerness to believe political leaders you agree with and to disbelieve those with whom you don't. Fight to force your mind down avenues and ways of thinking that you are unused to. Thinking outside established patterns is a radical act for the preservation of our collective freedom. Get good at it.

Learn to distinguish between real science and pseudoscience. In 2017, the Heartland Institute, a conservative think tank funded in part by the Mercer Family Foundation, sent beautifully produced textbooks on climate science to 300,000 schoolteachers across the United States. The book, originally targeting policy makers and published in 2015 to coincide with the Paris negotiations, was titled *Why Scientists Disagree About Global Warming* and began with this statement: 'Probably the most widely repeated claim in the debate over global warming is that "97% of scientists agree" that climate change is man-made and dangerous. This claim is not only false, but its presence in the debate is an insult to science.' This textbook, authored by 'distinguished climate scientists', was sent to teachers, with a letter urging them to use the book and its accompanying DVD in their classrooms. The Heartland Institute, which promotes denial of established climate science, encouraged people to 'seek out advice from independent, non-governmental organisations and scientists who are free of financial and political conflicts of interest' rather than relying on the UN Intergovernmental Panel on Climate Change (IPCC) for scientific advice.

It would have been extremely difficult for some recipients of that book to determine whether this was real science or bunk, and whether the authors were indeed distinguished climate scientists. In fact, one author was formerly director of environmental science at Peabody Energy (a coal company that went bankrupt). That author has a master's degree and a PhD in geography, not climate science. One of his credits is that he is the lead author of the reports of the Nongovernmental International Panel on Climate Change (NIPCC). Note the striking and confusing similarity of that name to the UN-backed IPCC. The NIPCC is actually a project sponsored by the Heartland Institute. Many teachers immediately saw the textbook as the unscientific propaganda it was, but those who didn't and used it in their classrooms had a lasting impact on their students.

This story teaches us a good lesson: even when a document looks 'official', is beautifully produced, and is authored by real scientists, we should approach its contents with caution. It is essential that you make the extra effort to determine whether what you are basing your opinions on is fact or fiction. Check where your information comes from.

If necessary, follow the money. Determine the source of the funding for the research in question, be it a climate science statement, report or article. See if the research is accredited by an established university or other well-known academic body. The simplest way to do this is to find out if the study was 'peer reviewed', meaning reviewed and evaluated by other experts in the field. For example, the IPCC report on 1.5 degrees Celsius, released in October 2018, was a collaboration of 91 authors and review editors from 40 different countries. Most mainstream newspapers will have an editorial policy to ensure that sources are either peer-reviewed or have similar criteria for reliability, but it is *always* worth checking.

Don't give up on climate deniers. As we enter the post-truth world more fully, the fault line between a desire for truth and an adherence to ideology runs closer to each of us. Some of us may have a natural inclination for one point of view but a deeper desire for truth, whereas others will exhibit a slavish adherence to one perspective, whatever the facts. In fact, those at the latter extreme have left the arena in which facts make a difference. Many people are now experiencing this even within their own families. Facts aren't enough to change the mind of a climate denier, so presenting statistics and sources won't help. If you reach them, it will be because you sincerely listened to them and strove to understand their concerns. By giving care, love and attention to every individual, we can counter the forces pulling us apart.

For people who came of age between the fall of the Berlin Wall and the fall of the Twin Towers, today's world can indeed appear strange. Those days were marked by a general consensus about how humanity should advance. Some may now wish for that simpler time, making us vulnerable to the promises of leaders who would take us back instead of focusing on what lies ahead.

The future will be different, it will be complex, and the genie of social media can't be put back in the bottle. There is no getting away from the fact that humanity needs to come to grips with the truth if it wishes to contain a monster of its own creation. If we wish to come together to address the climate crisis, and halt the rapidly accelerating extinctions that are now taking place in greater and greater numbers, we need to accept our responsibility to always defend the incontrovertible truths of climate change and their consequences. We are all responsible for what we hold to be true and for defending that truth against attack. We will succeed by applying a thoroughly critical approach to the information that shapes our ideas, opinions and actions. We will succeed by calling out falsehoods, particularly those that may determine how we act on climate change. Once this becomes a habit, once we become better practised at determining what is real, the fog of misinformation that we are currently cloaked in and the daily distractions vying for our attention will be easier to navigate. When we work this way to defend and advance a fact-based reality, the view of the regenerative future we want, and the path we will travel to get there, will come more sharply into focus.

Action 4

See Yourself as a Citizen –
Not as a Consumer

The South Indian monkey trap is an ingenious but cruel device. It consists of a coconut staked to the ground with a hole in it and a ball of sweet rice inside. A monkey approaches and fits his hand through the hole to grasp the rice he can smell inside. However, the hole is not large enough for his clenched fist to pass back through. His instinct is to keep his hand clasped over the ball of rice, so he is trapped by his instinct, not by anything physical: if he would let go of the rice, he would be free.

Such is our relationship with consumption (purchasing, using and throwing away): we know it is trapping us, but it has become so embedded in our psyche – to the point of being almost instinctive – that we cannot let go.

Much of what we buy is intended to enhance our sense of identity. Particular brands of clothes, soap, cookies, televisions, and cars are designed with a *tribe* in mind, their attributes carefully cultivated by the consumer goods companies that sell the products. Identity and consumption keep moving closer together. In the UK, for example, the average person consumes more than 65 pounds of clothes every year, equivalent to about five loads of laundry.[25] These purchases are driven mainly by the fact that fashion trends change each season. These cycles, by their very nature, require us to clear out our closets regularly and hop back in line for more clothes.

But the fashion industry has an enormous carbon footprint. Textile production is second only to the oil industry for pollution. It adds more greenhouse gases to our atmosphere than all international flights and maritime shipping combined. Estimates suggest that the fashion industry is responsible for a whopping 10 per cent of global CO_2 emissions,[26] and as we increase our consumption of fast fashion, the related emissions are set to grow rapidly.

Our engines of economic growth depend on us continuing to spend money. In the 1920s, some Americans were concerned that a new generation was emerging that had satisfied its needs, and that it would lead to a drag on growth. President Herbert Hoover's Committee on Recent Economic Change in 1929 concluded that advertising was necessary to create 'new wants that will make way for endlessly newer wants as fast as they are satisfied.'[27]

Today, consumer goods companies spend a great deal of money to make sure we remain stuck in the consumption cycle. Their marketing and advertising budgets are enormous. In the United States, the price of one 30-second advertisement during the Super Bowl – one of the most-watched sporting events on television – was more than $5 million in 2019.[28] Amazon, the online marketplace, raked in an extraordinary $10 billion in revenue from advertising sales in 2018 alone.[29] Every year more than $550 billion is spent on advertising in a world of consumption and fast consumerism.[30]

What is more, billions of products are intentionally designed to become obsolete, fuelling even more economic growth as we strive to replace them. Single-use plastics are the epitome of that, but obsolescence – the process of becoming outdated and discarded – is designed into almost all consumer goods. Warranties for certain products rarely go beyond three years because the product is likely to break after that period. And often a new item costs less than the replacement part. New software updates won't install on old computers, meaning those too must be replaced. The list is endless and depressing. As a result, the practice of mending, repairing and restoring is becoming a dying art.

In the global economy, supply chains often reach across the world and back again. Each link represents a different production stage, often performed by a different company, from the mining of precious metals in Bolivia for your smartphone to the packaging of the final product in China. As a result, it is hard to know which parts of the supply chains of major corporations practise sustainability and which contribute to climate change.

Here's what you can do.

Reclaim your idea of a good life. Consumerism is the prevailing definition of a good life: you are in perpetual pursuit of the almighty upgrade, whether it is to your phone, your clothes or your car. But rather than meeting our needs, buying things in order to achieve a sense of satisfaction or belonging can become addictive and lead to self-doubt and confusion about our very identity and life direction.[31] Identifying as a consumer – of any particular type of product or brand – implies passivity, and it also implies that consuming said product meets our needs.

Consumerism traps us into thinking we can purchase personality. Moreover, it eats up our mental space and creates a constricted view of the world, one in which our value and identity are built upon the proliferation of disposable waste. Psychological studies have shown that mass consumption creates a bigger and bigger hole in our lives that we keep trying to fill.[32] As we consciously or unconsciously attempt to consolidate our sense of identity through curated buying habits, we drive the engine of mass consumption faster and faster, bringing ourselves ever closer to the edge of disaster.

Despite all the ways culture is pushing us in the direction of blind consumerism, we can start to intentionally push back. We can develop the mental discipline to resist the imperatives of consumerism. We can change our consumption habits, and vote with our money for products that are sustainable.

Further, we can change the way we identify as consumers, to reboot our relationship with materialism. Freeing ourselves from the influence of advertising can be a liberating experience and a radical political act.

Become a better consumer. In the short term, we can improve matters by changing our consumption patterns within the system. Not all purchases are equal. Buying high-quality clothes made from organic cotton that will last and be handed down is different from buying cheap, disposable items that end up in landfill after a few weeks of wear. If you have the option of voting with your money, make more educated decisions about the products you do need to buy. Buy from companies that are public about their values, have made commitments to sustainability, and are part of organisations that certify they are following through on their pledges. The impact will be significant.

Vote with your money. Most important, eliminate waste. Apply the old-fashioned adage of reduce, reuse, recycle. When we need to buy things, our choices should be informed and enlightened.

Dematerialise. Consider how we made the change from vinyl, cassette tapes and CDs to downloading or streaming music. Technology in many instances now allows us to do without material objects while still enjoying the services that they provide. Less can be more. In the near future, even individual ownership of cars may cease to exist as the dominant paradigm – the transportation we need might be offered by shared vehicles, probably self-driving and certainly electric.[33] One day consumers may come to define themselves not as owners of products but as beneficiaries of systems of service delivery. Already the world's largest provider of overnight accommodation (Airbnb) owns no buildings. The world's largest provider of personal transport (Uber) owns no cars.[34] This shift from ownership to service provision will fundamentally change our relationship with consumerism. We can help accelerate it by engaging with it and welcoming it with open arms.

The story of the happy fisherman, first made popular by Paulo Coelho, has several versions. A content fisherman is relaxing on the beach in his little village after catching a few big fish. A businessman walks past, notices the bounty and asks the fisherman how long it took to catch all those big fish. Not very long, says the fisherman. The businessman asks why then, if it didn't take long, the fisherman doesn't spend more time at sea, so as to catch more fish. The fisherman replies that the fish he caught are enough to feed his whole family, and that when he finishes with his catch, he can go home to play with his children, take a nap with his wife, then join his friends for drinks and music making in the evening.

The businessman suggests to the fisherman that he could lend him some money to be more successful. Then the fisherman can spend more time at sea and buy a bigger boat to catch lots more fish that he could sell to make more money. He can then invest the money in more boats and set up a big fishing company. Over time the fishing company can go public on the stock exchange and make the fisherman millions.

'And then what?' asks the fisherman.

The businessman proudly explains that then the fisher-man can retire. He can finally enjoy spending his days as he wishes: catching a few fish in the morning, spending time playing with his children, taking an afternoon nap with his wife, and joining his friends for drinks and music making in the evenings.

It has been said that the most important things in life are not things. If, like Coelho's fisherman, we can learn to recognise what is *enough,* we might also move beyond the mindset of consumption and ownership, consciously avoiding the forces that feed that mindset. We can begin to appreciate that with a different approach to life, our capacity for happiness will increase and that our drain on the planet will dramatic-ally slow down.

Action 5
Move Beyond Fossil Fuels

The assumption that we will always need fossil fuels comes from incumbency thinking. In order to move beyond fossil fuels, we must let go of the conviction that they are *necessary* for humanity to thrive in the future. Only when this mindset is challenged can we migrate our thinking, finance, and infra-structure to the new energies.

Fossil fuel companies are deliberately slowing the trans-ition. As providers of these once plentiful and potent energy sources, the power of these companies has grown exponentially, and now their influence is deep and wide.

Many businesses continue to invest heavily in lobbying to water down new regulations that would help shift the economy beyond fossil fuels.[35] Some individuals in senior leadership positions, however, wish to address the issue and transform their businesses. That desire is sincere – we know this firsthand. But they are in a tough spot: if they shift their companies too far and too fast, they destabilise their business model, and investors will punish them. If they delay the shift too long, the value of their company may crumble. Several are playing the dangerous waiting game to be the 'last one out', continuing to derive income from the market space left by companies that are leaving fossil fuels behind.

Almost all governments are still subsidising fossil fuels. The fossil fuel industry may dispute it, but it receives huge government handouts. Globally, governments spend about $600 billion every year keeping prices of fossil fuels artificially low.[36] That's around three times as much as subsidies provided for renewable energy.[37] Governments may claim their administrations support renewable energy, but until they stop subsidising fossil fuels, our progress will stall.

Mark Carney, the governor of the Bank of England, famously said that unless we make a smooth transition from today's fossil-fuel-based economy to the fully decarbonised economy we need in the future, at some point there will be a 'jump to distress',[38] meaning that high-carbon assets will suddenly drop in value by a large percentage. Carney urged us to avoid it at all costs. When you think about how much of our economy is built on a foundation of fossil fuels, his prediction comes as no surprise. Entire industries, companies, and governments could go bankrupt or lose a lot of value very suddenly if we delay transition to the point of crisis.

If we allow a jump to distress to happen, it will affect all of us. Governments rely on tax receipts from fossil fuels to finance their services. Many pensions are invested in fossil fuels and in companies reliant on them. The systemic nature of the financial services system means that if a major drop in value occurs, it will quickly affect lots of other, seemingly unrelated entities. Such a jump to distress could make the financial crisis of 2008 pale in comparison.

Given all this, the urgent shift from fossil fuels must happen in a planned and measured way, and not as the result of panic. In 2017, heads of central banks came together to establish the Network for Greening the Financial System (NGFS) and are now united in their efforts to be vigilant of the impacts of climate change on global monetary stability.[39]

A growing body of financial research and information about how countries and companies are likely to perform in a future that is fundamentally different from the past is helping investors understand the risk. For example, Moody's rating agency (one of the highly influential agencies that assess risks to companies and countries) now has a controlling stake in RiskFirst, a firm that measures the physical risks of climate change.[40] Investors are reallocating capital away from what are now commonly known as 'stranded assets'. That reallocation is moving markets and catching the attention of corporate leaders, but it needs to go much further, much faster.

Stand up for 100 per cent renewable energy. In the past few years, energy generation from renewable sources has undergone an impressive surge. We are currently on track to supply 30 per cent of power demand in 2023 from renewables, and 50 per cent by 2030.[41] Corporations are taking the lead. Almost 200 companies, including well-known ones such as Apple, IKEA, Bank of America, Danone, eBay, Google, Mars, Nike, and Walmart, have already shifted to 100 per cent renewables as their source of electricity, or are on their way to doing so.[42] 75 per cent of people in Europe and North America support government taking strong action for electricity to be generated by 100 per cent renewable power.[43] To become our new reality, renewable power will have to be delivered at the systemic level by leaders in political and institutional situations of authority. Those leaders represent the priorities of the people who elect them, so let's vote for leaders who advocate clean energy.

If those in positions of power and influence today expect to be remembered as loyal public servants, responsible for representing the people, then they must look to the future with clearer vision. We should reward with our votes only the leaders who step forward with genuine insight.

We can do this with real confidence, because solar and wind power have developed at a speed and scale that few believed possible just a few years ago. With a 90 per cent drop in costs for solar panels in the past decade, renewables now compete with coal on price alone in most places around the world, and increasingly with gas as well.[44] A similar story is unfolding for both onshore and offshore wind energy production. The storage solutions required to smooth out energy from solar and wind are also rapidly evolving to become economically viable.

As costs have dropped, innovation is reimagining how energy grids of the future will operate. Far more intelligent and interconnected grids are emerging.

Make a time-bound, ambitious plan. We have ten years to cut our global emissions in half and another 20 years after that, at maximum, to get them to net-zero. Corporations and countries have great responsibility for leading the charge, but we can all play our part by reducing our own personal emissions. If we think clearly and act when we need to, there is enough time.[45] The 50 per cent reduction necessary over the next ten years is where we must now focus our attention. That is a global figure, but the number can be averaged out in this way: those of us who have been using far more than our share should reduce our emissions more than 50 per cent. Let's aim for a minimum of 60 per cent, knowing that we humans tend to overestimate what we can achieve in a year and underestimate what we can achieve in ten.

What would your life look like in ten years if you were using at least 60 per cent less fossil fuel than you are now?

Most of your current emissions probably come from flying, driving, and heating and cooling your house. The key culprits tend to be expensive items that we can't easily abandon, such as cars, boilers, and air conditioners. Once you have bought a car, you will use it, and while you may try to use it less, there is a limit to what you can achieve. Consider shifting to an electric vehicle within the next ten years. The increased efficiency and range of electric vehicles, combined with price drops and innovative financing models, are putting them within the reach of more and more of us. Even mid-range models are now capable of driving 150 miles in one stretch, and charging stations are more abundant than ever before.[46] Others may consider moving beyond the car, and even away from car ownership – a possibility that is becoming increasingly viable.

As for heating and cooling your house, you should aspire to buy renewable electricity through the grid and to generate more at home. Improving insulation and switching to electric heating all at once may seem daunting. Take one step at a time. Start by performing an energy audit in your home to identify energy leakages and inefficiencies. This will help you to prioritise your energy upgrade investments. You can do the cheaper energy improvements first, then plan phased investments over a few years when, say, a boiler would have to be replaced anyway. Over time you will save money and reduce emissions.

Reducing flying is likely to have the biggest impact if you live in a wealthy country. Much of what is wonderful about the world has come from the fact that we can visit different parts of it, have cultural exchanges and see amazing places. It is an unbelievable privilege for those who are able to afford it to get on a plane in one part of the world and get off, ten hours later, on the other side. If you enjoy travel adventures, take business trips or visit family abroad, you will not find it easy to give up flying.

Only 6 per cent of the world's population has ever set foot on a plane.[47] If you are among them, it is incumbent upon you to take a stance and make a plan. You might decide never to set foot on a plane again, and if you do, we applaud and celebrate you. But in reality, that may not be possible for you today, but you can still make a contribution. You can commit to not flying for holidays, or to taking the train to places within, say, 500 miles of your home. You might commit to taking only a certain number of flights per year, or to taking meetings via video conferencing.

However you approach it, air transportation is one of the critical issues we are going to have to grapple with on the path to a 60 per cent reduction by 2030. Neither it nor the other changes discussed here have to be frightening. When people consider such lifestyle changes, they can become alarmed and feel that something precious is being taken from them. However the opposite is the case. While we may resist change, the reality is that the speed, scale and reckless use of resources in our wasteful economy are making few of us happy. As we focus on making thoughtful changes to help preserve what we really care about, finding a sense of purpose often improves our quality of life. Try it for yourself, and see what you find.

Action 6
Reforest the Earth

The future we must choose will require us to pay more attention to our bond with nature. Ancient stands of trees teeming with life are integral to our survival. Extracting more and more output from an increasingly depleted and exhausted soil is a formula for our own destruction. If we want to thrive over the long term, we need to find the sweet spot of working to regenerate nature for its own benefit and ours, and drawing from it only what we need to support our lives. Achieving this balance on a global scale is still possible. We can be the generation to achieve it.

Forests create the conditions for forests, in a self-sustaining system. They give up moisture to the sky, which creates clouds and rain, moving water back to all parts of the forest. Microscopic fungi in vast underground networks of mycelia stretch between trees across thousands of miles and connect them, sharing nutrients. Soils build up and create the rich foundation for future generations of trees. This symbiotic interplay makes a forest vulnerable, however. If we destroy enough of it, or fragment it, thereby hindering its interconnectedness, the whole system can collapse. We will lose the great forests of this Earth the way, in an old saying, people go bankrupt: first very slowly, and then very fast.

Since the dawn of agriculture, humans have cut down approximately 3 trillion trees, or half the trees on Earth. As a result, almost half the land on our planet has been severely degraded from its natural state. In 2018 alone, 12 million hectares of forest – equivalent to 30 football fields a minute – were razed, a third of which was pristine primary rainforest.[48] If we carry on in the same vein, we will destroy everything that is left of our forests within a very few short decades. Even if we avert this fate, generations to come will wonder in astonishment at how close we came and how mindlessly we almost threw the forests away.

Almost all tropical deforestation is driven by demand for four commodities: beef, soy, palm oil, and wood. Beef cattle are responsible for more than double the deforestation of the other three combined. In the Amazon, providing land for beef cattle to graze on is directly responsible for more than 80 per cent of the deforestation.[49] In addition, much of the soy is used as feedstock for chickens, pigs and cattle. This situation is bad and about to get worse, with Brazil lifting previous forest-protecting policies,[50] and China now massively increasing its meat and dairy consumption.[51]

Industrial agriculture and the food industry, which often prioritise profitable food over nutritious food, are almost as big a driver of climate change as fossil fuels. Yet much of the food produced is never eaten. It doesn't even necessarily get to the people who need it. In the Global South, a lack of roads and storage facilities means that food often rots before it gets to people, and even if it does reach them in time, they might not have the money to buy it. In the Global North, food languishes in home and supermarket refrigerators until well past its use-by date, or it is left uneaten on the plate at the end of a meal and then thrown away. Such waste then drives greater food production.

We can achieve food security for all. At least two distinguished ecologists have calculated we could feed the world adequately by making selective improvements in agricultural productivity, sharply reducing food waste and changing our diets,[52] which health experts recommend anyway.[53] We can do all these things without destroying another square inch of nature.

Plant trees. Vast land areas around the world are potentially available for reforestation and tree planting. One study found that 900 million hectares, about the size of the entire United States, are available for reforestation without interfering with either human habitation or agriculture.[54] Once new forests were mature, they would absorb and store 205 billion tons of carbon, while supporting biodiversity and making the planet more beautiful. That equates to absorption of nearly 70 per cent of all the CO_2 released into the atmosphere since the Industrial Revolution.

In addressing climate change, few actions are as critical, as urgent, or as simple as planting trees. This ancient carbon-absorbing technology needs no high technology, is completely safe and is very cheap. It literally reverses the process that has led to climate change, in that as trees (and all other biomass) grow, they absorb CO_2 from the air, release oxygen and return carbon to its rightful location: in the soil. In addition, trees provide coveted green areas in cities, reduce ambient temperature, may produce food, and stabilise aquifers in rural and suburban areas.

Unfortunately, over the past five to ten years, we have come to think of planting trees and reforesting as a penance we must pay for the sin of emitting greenhouse gases, or worse yet, as a pretended benefit that hides the reality of emissions. 'Offsetting' has developed a bad reputation among some environmentalists. It is time to correct this mistake. Every single one of us should plant one tree, ten trees or 20. Don't even think of it as an offset – in itself it is a critically important contribution to addressing climate change now, without the need for sophisticated energy technologies. Those will be developed, but even when we count on them, we will still need to absorb carbon out of the air to reach net-zero emissions.

In short, we could return the climate to how it was decades ago just by planting trees.[55]

Massive reforestation and restoration provide real benefits for people. In China in the 1990s, vast areas of land began to resemble the Dust Bowl of the American Midwest, but China was able to halt this rapid degradation. Programmes were established to reforest 100 million hectares by paying farmers directly to plant trees. The programme is ongoing and highly successful. It has resulted in more stable rainfall, more fertile soil and increased production from farmland.[56] Ethiopia, having diminished its forest cover to a mere 4 per cent of its territory, undertook a record-breaking campaign by planting 350 million trees at 1,000 sites across the country, most of which were planted in a single day.[57] Not all of them will survive, but those that do make an important contribution.

The benefits of planting trees are not limited to rural or agricultural areas. Trees will cool a city by up to 50 degrees Fahrenheit (10 degrees Celsius).[58] That amount can make up for the significant additional heat that cities will have to endure under any climate scenario, and as cities in India are already reaching temperatures in excess of 122 degrees Fahrenheit (50 degrees Celsius), it could mean the difference between life and death for millions of people. Trees also clean the air in cities by filtering fine particulate matter and absorbing pollutants. They regulate water flow, buffer flooding and increase urban biodiversity. Their impact is so pronounced that urban properties surrounded by trees are worth an average of 20 per cent more than those that are not.[59] If we are to make the transition to urban living that is needed to provide space for nature to thrive, we need to bring nature into cities and integrate it as never before.

Let nature flourish. The term *rewilding* has been coined to describe the growing practice of allowing land to return to its natural processes. Rewilding has the potential to radically change the carbon balance of the atmosphere and to preserve the web of life. Multiple large- and small-scale rewilding initiatives are already taking place all over the world. An excellent example is the Knepp Wildland Project in West Sussex, England. In 2001, the project obtained more than 3,500 acres of land that had been farmed intensively since the Second World War. The land was severely degraded, and the farm had rarely made a profit. Knepp Wildland's ethos is to allow natural processes to play out rather than aiming for any particular goals or outcomes. Free-roaming grazing animals – cattle, ponies, pigs, and deer – drive this process-led regeneration, acting as proxies for herbivores that would have grazed the land thousands of years ago. Their different grazing preferences create a mosaic of habitats from grassland and scrub to open-grown trees and wood pasture. These animals need minimal intervention. At low cost, they provide wild-range, slow-grown, pasture-fed organic meat for which the market is growing. In just over a decade, Knepp has seen astonishing results in biodiversity. It is now a breeding hotspot for purple emperor butterflies, turtle doves and 2 per cent of the UK's population of nightingales.

Go plant-based. If you eat less meat and dairy, your carbon footprint will decrease, and your health will improve. Eating less meat and dairy is better, and eating none at all is best. While this may feel like a stretch for most of us, for the vast majority of human history we ate very little meat.[60]

Many countries are already shifting towards plant-based diets. Even if you feel that you cannot completely forgo meat and dairy, adopting a flexible diet in which you enjoy other foods for certain meals or certain days of the week can have a huge impact. In reality, this is likely to be where the biggest dietary changes will come in the next years. In many countries the number of people planning to become vegan or vegetarian is relatively low, but fully 50 per cent of the US population would like to eat *less* meat. Plant-based meat replacements are already becoming cheaper, more efficient and more delicious. By 2040, these products are expected to make up 60 per cent of the market, up from 10 per cent today.[61] The market is beginning to recognise the future of plant-based food. You have the chance to join a food revolution by adopting and normalising a more plant-based diet.

Boycott products contributing to deforestation. Too many ingredients in the products we consume every day come from deforested land. In 2010, Greenpeace released an advertisement featuring an office worker opening a Kit Kat. However, the bar was made not of chocolate but of orangutan fingers, and as the office worker took a bite, blood poured across his keyboard.[62] The video hit a nerve, helping people make the connection between confectionary ingredients and the mass destruction of the orangutan's natural habitat. More than 200,000 emails were sent to Nestlé; protests were held outside its offices. Within six weeks one of the largest companies in the world completely reversed its policy, committing to zero-deforestation palm oil.[63]

It's easy to forget how much power we all have if we

choose to use it. If a company is engaging in destructive land practices, we can work to make that fact clear to everyone. As that happens, you can remove your consent from that company by refusing to buy its products.

We are not powerless.

Action 7
Invest in a Clean Economy

A linear model of growth rewards extraction and pollution. We need to move from that model toward one that regenerates natural systems. We are going to require a clean economy that operates in harmony with nature, repurposes used resources as much as possible, minimises waste and actively replenishes depleted resources.

This new economic model will need better policies and strong institutions so that the great market forces of investment and entrepreneurialism can work towards regeneration instead of extraction. Finance and investment will play a key role. While we have managed capitalism moderately well over the centuries, with successful institutions such as law, taxation and charity, we have not yet perfected it. Now is the time to do so.

We are used to thinking of the economy as the primary indicator of how we are performing as a species. More economic growth is good, less is bad; negative growth, or a recession, is a disaster. Politicians will do anything in their power to keep the numbers moving in the upwards direction, and most regard this as their principal responsibility.

Economic growth is currently measured by GDP (gross domestic product), the market value of goods and services produced in a year. The idea that endless GDP growth is the aim of responsible countries is highly embedded into our cultures and becomes self-perpetuating, as the media, politicians, business leaders and others constantly refer to it as second nature.[64]

But GDP is a poor marker of what human beings need in order to thrive, as it is all about extracting, using and discarding resources. As a marker of success, it does not effectively take into consideration the impacts of pollution or inequality, or prioritise the value of health, education, or even happiness. It also places no value on the actions that regenerate degraded lands or that bring ailing oceans back to health. To illustrate the point, if you drink coffee from a disposable cup every day, GDP will go up, but the forests will disappear and emissions will go up too. If you drink coffee from a reusable ceramic mug, GDP will go down. If you throw away your ceramic mug every day and buy a new one, GDP will go through the roof.

In the current transition, strictly linear GDP growth can no longer be the priority. More stuff does not mean a better life, and indeed it is contributing to our existential crisis. Moving away from quantity of products that can be purchased, we must reorient our underlying sense of value towards quality of life, including within all of Earth's ecosystems. Prioritising growth according to its contribution to the Sustainable Development Goals (SDGs) would be a good place to start. These 17 interconnected goals aspire to sustainably increase global prosperity, equality and well-being.[65]

Put your money where it matters. The capital tends to flow towards investments that have worked in the past, as if the future will resemble the past in any meaningful way. The world's capital is guarded by ranks of extremely cautious people who are looking to secure a good return, and their top priority is often to avoid risking a loss of value. This is technically right, of course, but it presents us with a problem. We're not going to create the future we want without some risk.

In June 2019, the Norwegian parliament voted into law new plans for its sovereign wealth fund (the world's largest, managing $1 trillion in assets). It will divest more than $13 billion of investments in fossil fuels and invest up to $20 billion in renewables, beginning with wind and solar projects in developed markets.[66]

You can help precipitate similar seismic shifts in allocation of capital. In 2012, Bill McKibben and 350.org began a grassroots divestment campaign to encourage financial institutions to stop investing in projects and companies that perpetuate the causes of climate change.[67] It has grown into one of the most successful campaigns in history. Financial firms with more than $8 trillion in combined assets have divested their fossil fuel holdings. This has made money available for climate solutions and sent a warning signal to those still building the past. In 2016, Peabody, the world's largest coal company, listed divestment as one of the reasons for its bankruptcy.[68] Shell has listed divestment as a material risk to the future of its business.[69]

Divesting from the past and reinvesting in the future can be done right now. Your money has the power to destroy or to build, and it is no longer acceptable to remain oblivious to the fact. If you have a pension fund or savings, find out where your money is invested. Do not underestimate the power of the default option in defined pension schemes – if you work for a company that has such a scheme, request that it shift away from fossil fuels. Write to your pension fund trustees and find out if they are divesting from the old economy or how they propose to change the behaviour of corporations they are invested in so as to promote the clean economy. Encourage your friends and colleagues to do the same.

Once capital starts flowing in increasing amounts to companies and projects that are advancing the future – and we are making serious progress in that direction already – a moment will come when we reach the zenith of our uphill efforts and things will start to roll more easily in the right direction. We are already seeing that dirty, polluting, irresponsible investments perform less well than the alternatives. Companies that shy away from considering the future of the planet are also getting awkward questions from customers (keep asking them!) and investors, and are struggling to find bright young people to work for them. With continued pressure, the money and momentum will start flowing to those who are building the clean economy.

The building blocks for a regenerative economy are already robust and thriving around the world. In January 2019, Jacinda Ardern, Prime Minister of New Zealand, announced that her government would soon present a 'well-being budget' to gauge the long-term impact of policy on the quality of people's lives. 'We need to address the societal well-being of our nation, not just the economic well-being', she said. This type of thinking, Prime Minister Ardern argued, could help us shift beyond short-term cycles and learn to see politics through a lens of 'kindness, empathy and well-being'.[70] This is what we are called to do, as we work to build the infrastructure and systems that will benefit us, and retire those that are harming us.

Economic growth can deliver tremendous benefits, and economic growth has lifted more people out of poverty than any other model in history. But the days of valuing how quickly we can dig stuff up and turn it into rubbish have to come to an end, not as a matter of ideology or policy but as a matter of survival. The reduction of poverty under the old model may well be temporary, since our structure of prioritising short-termism and GDP will likely send many people back into unforgiving poverty as climate change accelerates. The good news is that economists increasingly consider the 17 Sustainable Development Goals to be sensible objectives.

Advancing the SDG framework makes it absolutely possible for us to achieve sustainable growth, effect emissions reductions, and reduce poverty in consonance with one another in mutually reinforcing systems.

In Costa Rica, President José Figueres Ferrer, Christiana's father, made the decision in 1948 to abolish the army. He invested in education and expanded forest cover from a low of less than 20 per cent. Now Costa Rica has one of the highest literacy rates in Latin America,[71] forest cover is more than 50 per cent,[72] and the nation's electricity is provided almost exclusively by renewable energy. Costa Rica measures its progress both by GDP and by indicators that help the government make decisions that maximise well-being. On the Happy Planet Index, Costa Rica ranked number one as the happiest place on Earth in 2009, 2012 and 2018.[73]

Action 8
Use Technology Responsibly

Evolving new technologies have enormous potential for delivering emissions reductions. We must embrace them carefully but rapidly and not rely on them as a silver bullet. As we grow more comfortable with machines being part of our lives, we will need to use technology responsibly, mindful of its power and influence, and ensure that proper governance systems are in place.

If we make it through the climate crisis and arrive on the other side with humanity and the planet intact, it will be largely because we have learned to live well with technology.

Artificial intelligence (AI) supported by sensors (to gather data) and robotics (to automate physical activities) together with the network of smart devices known as the 'internet of things' have huge potential to become our greatest allies in the fight for survival.[74] But these very same technologies are also the ones that could destroy that better future. For example, autonomous self-driving electric vehicles could eliminate the need for unnecessary private ownership of vehicles, but on the downside, they could also allow unscrupulous governing bodies to track and control the movements of every citizen.

A fire that warms you on a cold night is good; one that consumes your home is bad.

Likewise, technology is neither inherently good nor inherently bad. It just has to be managed properly.

Many people alive today will at some point likely encounter a machine that is smarter than they are in almost every way. The world famously got a taste of what that might be like in 2017. The AI Programme AlphaGo Zero figured out how to win at the ancient and notoriously difficult Chinese strategy game of Go, learning entirely by itself, essentially accumulating thousands of years of human knowledge, and improving on it, in just 40 days.[75]

Deep Mind, the company that developed AlphaGo Zero, says the technology is not limited to machines that can outcompete human beings in strategy games but is intended to be used to inform new technology that will positively impact society.[76] But we can't rely on the promises of corporations to ensure that a technology is aligned with our goals for regenerating nature and pursuing the conditions that will help humanity thrive.

AI machines learn quickly, although we may not be able to predict exactly what they will be used for. Machines could become better at extracting what resources remain on Earth and hoarding them for those who control the technology – which is why protection against the abuse of AI needs to be woven into policy oversight and governance from the start.

Politicians and CEOs who are unwilling to lead or do what we need to confront the climate crisis have often touted future technology as a solution. But if we allow the potential of future technology to blind us to the scale and urgency of what we need to do today, we will be taking a terrible risk. Not only might innovations not arrive in time, but new technology will fit well only into a society that is already moving in the right direction. Belief in innovation is no excuse for lack of a plan.

To be sure, we need technology to avert climate disaster, but technology also has huge potential to increase the already-vast wealth disparities in our societies. In a world where 70 per cent of the population has to survive on a share of only 2.5 per cent of global wealth,[77] the rise of automation could exacerbate inequality and social instability, and complicate the advance of solutions to complex problems like climate change.

For all the talk in certain political circles about immigration taking jobs away from native citizens, it is automation that is driving the vast majority of job losses around the world.[78] This problem will worsen in coming decades. Likewise, the decline of meat consumption, as it is replaced with plant-based and lab-grown alternatives, will transform the economies of whole countries. In Brazil, more than 20 million people are involved in the agriculture industry.[79] Up to two-thirds of them grow either cattle for beef, or soy to feed cattle. To switch to more sustainable agriculture, they could convert the land to biofuel production, assuming increased demand for such in the near future. The shift away from beef and towards advanced biofuels will have huge benefits ecologically, but if the transition is managed badly, without supporting alternative training or jobs, the sudden unemployment of millions could result in enormous human hardship, increasing the appeal of extremist politicians. Even if we develop all the technology needed to address the climate crisis, humans may be so impacted by the transition that we will elect leaders who pander to populist impulses and divert our focus from the narrow gate towards a regenerative future.

If properly managed, machines might make all the difference in our ability to deal with the climate crisis in time. Almost every sector that requires breakthroughs to bring about a regenerative future will be massively aided by machine learning. For example, one of the big problems associated with securing large amounts of renewable power on energy grids is that its generation is intermittent – producing only when the sun is shining or the wind is blowing.

With AI algorithms, it is now possible to completely r edesign our centralised energy grids. AI-informed energy grids can be much more decentralised, acting as neural networks, dynamically predicting what power is needed when. AI-informed grids would 'intuitively' map supply and demand, flexing between storage and energy flow so that greater amounts of renewable energy can be produced, thus reducing gas and coal use, perhaps completely.[80]

AI is accelerating our decarbonisation efforts in many other areas. Machine learning is being used to prevent the leakage of methane from gas pipelines, to accelerate the development of solar fuels (synthetic chemical fuels produced directly/indirectly from solar energy), to improve battery storage technologies, to optimise freight and transport for better efficiency, to reduce energy use in buildings, to plant forests using drones and much more.[81] AI is also showing promising signs of improving our ability to predict extreme weather and even of removing greenhouse gases directly from the air.

Reaching the Paris Agreement was complicated, but agreeing on a collective global approach to governing AI could be even more so. Right now countries are in a race to develop the skills and conditions to be leaders in this new field, and different populations have different attitudes about the acceptable degree of involvement of AI in our lives. For instance, people in Nigeria and Turkey would be happy to have AI systems perform major surgery on them, but people in Germany and Belgium would not.[82] Governments experienced different degrees of pressure to develop appropriate guidelines for managing AI, and as a result some are very lax and some are highly stringent.[83]

Understandable as this is, it isn't really good enough for something as important as dealing with the climate crisis. The effort of the French and Canadian governments to create an International Panel for Artificial Intelligence is a good start.[84]

Find out if your government, your local community or the company you work for is investing in AI, and what they are using it for. Take responsibility for pressuring them, in whatever way you can, to look to the international efforts already under way, and to put policies in place to ensure that the AI they support will also accelerate the regenerative future, not hinder our chances of success.

In a few decades more than 9 billion people could inhabit the planet, possibly more than 10 billion. It will be impossible for so many people to live here if we have the same impact per capita on our atmosphere as we do today. Technology, specifically machine learning and AI, has the potential to transform our presence here. Issues and problems, including how we can effectively use natural resources in a circular rather than linear way, that have long eluded us may finally be unlocked.

When AlphaGo Zero was learning to play and win at Go, the developers noticed that as it taught itself techniques perfected by professional players over generations, it occasionally made decisions to discard those techniques in favour of new, better ones that human beings had not yet had time to learn. In a race against time, the speed of learning that AI offers has extraordinary – exponential – potential to accelerate climate solutions, if it is deployed and governed well.

A humbling story of how this might unfold took place at Google's data centres in 2016. For more than ten years, Google engineers had been at the cutting edge of optimising their data systems. Their servers were among the most efficient in the world, and it seemed that any improvements from then on would be marginal. Then they unleashed DeepMind algorithms on the system. Energy demand for cooling was consistently reduced by 40 per cent.[85] This illustration is just a tiny example of the power of AI to make possible what seems impossible to the human mind.

At present, investment in applying AI to the climate crisis is lower than it should be. In the future, governments and corporations around the world will have to carefully support the responsible application of AI and invest quickly in its capacity to deliver material breakthroughs in emissions reductions. In that scenario, technology may be our greatest ally on the road to a brighter future.

Action 9
Build Gender Equality

We must ensure that decision-making at all levels of society involves increasing numbers of women, because when women lead, good things happen. That is the unequivocal conclusion of years of research. Women often have a leadership style that makes them more open and sensitive to a wide range of views, and they are better at working collaboratively, with a longer-term perspective. These traits are essential to responding to the climate crisis.[86]

We know this because the early evidence is already in. Companies, countries, NGOs and financial institutions all take stronger climate action when they are led by women or have a high proportion of women in decision-making roles.[87] Recasting our society so that women play at least an equal role in decision-making at all levels (family, community, professions, government) is now a matter of survival.

In many countries, discrimination based on gender is assumed to be a thing of the past. Yet studies show that all industries still strongly tend to overestimate male performance and underestimate female performance. While women are aware of this discrepancy, men tend to dismiss it. The vast majority of leadership role models remain male: just look at any photo of G20 leaders from any year. The well-publicised pay gap (women are paid 20 per cent less than men for the same work) is another manifestation and shows that many perceptions continue to be subjective and discriminatory.[88]

Before we can work to correct the imbalance of power and decision-making, we have to acknowledge that it exists, often but not always based on structural unconscious bias. Right now that is still lost on many.

Nonetheless many women have recognised the unique gravity of our situation on climate change. Intrepid leaders like Natalie Isaacs, Isra Hirsi, Nakabuye Flavia, Greta Thunberg, and Penelope Lea have mobilised millions of young people who are now demanding urgent climate action and implementing it themselves. Women are at the forefront of collaborative efforts to support each other in the face of our changing climate. In many countries, women's intimate knowledge of the land means they are quicker to spot environmental changes, to learn from them and, out of necessity, find ways to adapt. Women are pioneers of innovative climate solutions within their communities, and they are instinctively good at deep listening, at empathy and collective wisdom gathering, especially in times of transition. These qualities have never been more important or necessary.

A world with true gender equality would look different from ours. Some seem to assume that it would look the same but with a tilted gender power balance. But the interesting element of gender equality, apart from its evident moral rightness, is the opportunity it provides for all of humanity to co-create a world that is regenerative and in which we can thrive together. Nations with greater female representation in positions of power have smaller climate footprints. Companies with women on their executive boards are far more likely to invest in renewable energy and develop products that help solve the climate crisis. Women legislators vote for environmental protections almost twice as frequently as men, and women who lead investment firms are twice as likely to make investment decisions based on how companies treat their employees and the environment.[89]

It is imperative that women be afforded educational opportunities worldwide. Educated women can work, be economically more productive and help society make better decisions. Crucially, education helps women stand up for themselves and empowers them to make their own choices, in particular about their reproductive health. Keeping girls in school means they are less likely to marry young or have as many children. According to the Brookings Institution, in certain parts of the world, a girl with 12 years of education compared to one with no schooling will have up to five fewer children in her lifetime.[90]

Today, 130 million girls are being denied the right to attend school, condemning a massive number of future women to constant pregnancy, bringing more and more children into parts of the world that will scarcely be able to support them. By these calculations, 100 per cent enrollment of girls in school today would lessen the anticipated global population in 2050 by 843 million people,[91] a boon in confronting the climate crisis.

If you are a woman, now is the time to consider running

for public office or being more assertive about a deserved promotion at work. If you are a man, now is the time to support and encourage your female colleagues, partners, friends and family members. Women may feel particularly empowered by joining a wider movement or a cohort that shares their aims. The Brand New Congress movement in the United States, which played a significant role in a record number of women being chosen for the 2018 primaries, is a powerful example.[92] Female candidates, including Alexandria Ocasio-Cortez – now a seriously influential leader on climate action – drew on huge reserves of confidence to run for office by standing shoulder to shoulder with other women.[93]

We will be able to manage climate change better if we can improve the ratio of women making the decisions about how to do it. It's time to either become one of those decision-makers or support women you know to become one.

In the remote, sun-cracked desert of India's westernmost state, Gujarat, women are harnessing renewable power and improving their livelihoods by acting collectively. Gujarat, the source of nearly 76 per cent of India's salt, remains largely disconnected from the electrical grid. For decades, more than 40,000 salt pan worker families (locally called *agariyas*) have relied on diesel-powered pumps – often spending more than 40 per cent of their annual revenue for the season's production. Now that is all changing. With visionary leadership and support from Reemaben Nanavaty, a native of Gujarat and director of the Self Employed Women's Association (SEWA)

– which, with 2 million members, is the largest trade union for informal workers in the world – the *agariyas* are shifting to solar. The first 1,000 women who made the shift have benefited from a doubling of their income – helping them to achieve greater financial and social independence, and enabling them to send their children to middle and high school. When rolled out to the 15,000 SEWA members who work on the salt pans, the project will prevent the emissions of 115,000 metric tons of carbon dioxide – the equivalent of taking nearly 25,000 cars off the road.[94]

Solar Sister, a social enterprise operating in Nigeria and Tanzania, recruits women and trains them to sell affordable, renewable energy sources, like solar lamps and clean cookstoves. Deforestation and climate change mean women must often walk farther than they used to in order to collect water or find firewood for cooking. If they don't collect enough water or firewood, they are more likely to experience domestic violence. The increased workload also means that they have less time to spend on education or income-generating activities. Solar Sister has recruited and trained 4,000 women who are now entrepreneurs and have brought clean energy solutions to 1.6 million people in Africa and relieved some of the pressure on women.[95]

These are just two examples of women improving their own lives and livelihoods and those of their sisters when given the resources and freedom they need.

The potential is global.

Action 10
Engage in Politics

Finally, the action that we feel is ultimately the most important. Democracies are threatened by the climate crisis and
must evolve to meet the challenge. In order to help them do
so, we all need to actively participate.

The transition to a regenerative world is possible only if
we have stable political systems that are responsive to our
planet's changing needs and our citizens' changing desires.
Since climate change threatens political security itself,[96]
stability is both an essential condition for the transition and
an outcome of managing it successfully.

If the first duty of government is to protect its people,
then across much of the world the form of democracy we
have become used to is failing. Climate change is an existential threat and is likely to intensify faster than most people
today realise. If our systems of government can't protect us
from that existential threat, they will in time be replaced.
But those replacements may take a long time to evolve and
will not necessarily be any better at advancing us towards a
regenerative future in the available time frame.

In many countries today, corporate interests have captured our democracies. Just as with the tobacco industry, a
small minority of companies have used a relatively limited
amount of money to purchase extraordinary influence in
major legislative capitals and thereby have prevented elected

representatives from protecting the people. Often this occurs through trade associations, so even when corporations themselves do not directly lobby for an outcome, it is done on their behalf.[97]

This has become a major issue. In the United States, for example, in 2016, the National Association of Manufacturers (NAM) won a long-fought battle to delay implementation of the Clean Power Plan. In 2017, NAM supported the US withdrawal from the Paris Agreement. Companies such as Microsoft, Procter & Gamble, Corning, and Intel are all members of NAM, yet all claim to support strong climate action under the Paris Agreement.[98]

On a national level, voter action (or inaction) and intent underpin larger global moves. Over the last 20 years, climate change has been steadily climbing up the list of voter priorities.[99] While this is good news, no significant proportion of voters actually see climate as their highest priority. That is a serious problem. In the United States, new presidents have a very short window of time to actually get big things done. For example, Barack Obama came into office very committed to taking strong action on climate, and he had a majority in both houses of Congress. He could have chosen to prioritise – and would probably have passed – ambitious climate legislation. However, instead he made a decision to pursue heath care reform, another campaign pledge and a domestic priority. Passing heath care required Obama to use up a significant part of his political capital, and it built a knot of fierce resistance to his other policies in the Republican

Party, to the point that they stonewalled anything he proposed. As a result, not until his second term was he able to turn his political attention to climate change. Even then, it was only by using executive power that he made progress, not through legislation.

Rather than wait for things to get worse, we must embrace engagement at all levels of politics. We must see it as one of our most pressing responsibilities, and we must hold every politician to account. We must elect only leaders who see far-reaching action on climate change as their absolute first priority and who are prepared to act on the first day they assume office. Large numbers of people *must vote on climate change* as their number-one priority. As we are in the midst of the most dire emergency, we must urgently demand that those who seek high office offer solutions commensurate with the scale of the problem. Their policy platforms must strictly be informed by science.

It's time to participate in non-violent political movements wherever possible.

In April 2019, the group Extinction Rebellion, building on years of work by various non-profit organisations, some politicians, and other activists, seized the moment and began a series of global protests, the first of which was to take over central London for ten days in non-violent protest. Thousands of first-time activists, people who had never marched or signed a petition in their lives, blocked roads, linked arms, and planted trees on Waterloo Bridge. Within two months of that initial protest, the UK declared a climate emergency, adopted a target of net-zero emissions by 2050 (less ambitious than what Extinction Rebellion was calling for, but still a big step), and established a citizens' assembly to look at how it could be achieved.[100]

Civil resistance by members of the public can outdo efforts by political elites to achieve radical change. This is not an aberration; it is how change happens, typically when injustice in the prevailing system becomes too great.

Civil disobedience is not only a moral choice, it is also the most powerful way of shaping world politics.[101] Historically, systemic political shifts have required civil disobedience on a significant scale. Few have occurred without it. The numbers required may seem large, but they are not impossible. History has shown that when approximately 3.5 per cent of the population participates in non-violent protest, success becomes inevitable.[102] No non-violent protest has ever failed to achieve its aims once it reached that threshold of participation. In the UK, this would be 2.3 million people. In the United States, 11 million.

These numbers are now within our reach.

The remarkable rise to prominence of Greta Thunberg and the Fridays for Future movement is showing us that the world is ready for the next phase of direct action.[103] Greta's single, defiant act of civil disobedience – striking from school every Friday – has captured the zeitgeist. She inspired, in a relatively short period of time, a peaceful process for igniting and harnessing the anger of millions of young people in many countries and enrolled them in regular climate activism.

Adding further momentum to the successful capital divestment movement (in which money is moving away from assets linked to fossil fuels), in 2019, the head of the Organisation of Petroleum Exporting Countries (OPEC) described the mass mobilisation of world opinion against oil as the greatest threat its industry faces.[104] This mobilisation has as its motive force people from all walks of life, spanning all generations, across all continents. Every additional person who chooses to participate will bring us closer to the tipping point for success.

We acknowledge that participating in school strikes or civil disobedience demonstrations is not always possible or, in undemocratic societies and even in some democracies around the world, safe. What is important is for you to assess the avenues that might be open to you to engage in the political process and to find ways to work within them.

Beyond directly addressing governments, other political actions are needed. Corporations and trade associations fund and engage in political lobbying against citizen action on climate change. We need to remove our consent from these corporations. The simplest way is to vote with your money: stop buying their stocks, and stop buying their products and services where alternatives exist. Talk to your bank, talk to the institutions that manage your insurance products or debts. Find out if your money is invested in these corporations and ask for alternative options. Some financial institutions are already taking protective action, but others may not yet feel sufficient pressure from their customers to make a serious shift in capital allocation.

Governments that are stable now and trying to find ways to meet this challenge should be worked with, not dismantled. We all have a responsibility to exert what leverage we can inside the traditional power systems and push them as far and fast as we can. As we press both inside and outside the system for the overdue political changes that need to occur, we should also be mindful of the role that institutions have played in upholding our basic rights and our ability to weather transitions together. For hundreds of years – thousands, in some cases – our institutions of government, learning, communication, law, and religion have held us to a norm. It is possible to argue that this is what has kept us back, and at times in history that has been true. But equally true is that they have protected us from our worst instincts at moments of rage and insanity. Let's be mindful of what they have given us and find ways, when appropriate, to protect them. Once they are gone, they cannot be easily replaced.

Because climate change is unlike any other challenge that humanity has had to face, we have no template for the kind of political, economic and societal transformation needed now – but there are a range of extraordinary examples we can learn from. Movements of civil disobedience from the early twentieth-century suffragettes to Gandhi's drive for Indian independence to Martin Luther King, Jr., and the 1960s civil rights movement to the 2003 Rose Revolution in Georgia – to name just a few – are all inspirational insofar as they mobilised vast numbers of people to champion their causes. An open, inclusive narrative and a sense of working collectively to change history for the better took them further than they ever imagined possible. As Nelson Mandela said, 'It always seems impossible until it is done.'

Now is the time for us to participate – in our schools, businesses, communities, towns and countries – to ensure that the battle to survive the climate crisis becomes the biggest political movement in history. It is not about changing governments or political leaders. It is about waging sustained political action and engagement. The ingredients to achieve our goal are ripe. We have huge momentum with millions of people on the streets calling for change. Corporations, cities, investors and governments all over the world are taking highly sophisticated and coordinated action towards a 1.5-degree-Celsius future, and are open and listening to the calls of emergency from the streets.

If democracy is to survive and thrive into the twenty-first century, climate change is the one big test that it cannot fail.

A New Story

We want you to know two things.

First, even at this late hour we still have a choice about our future, and therefore every action we take from this moment forward counts.

Second, we are capable of making the right choices about our own destiny. We are not doomed to a devastating future, and humanity is not flawed and incapable of responding to big problems if we act.

Future generations will most likely look back at this moment as the single most significant turning point for action.

But the path we have set out is not easy, and success is not assured. The road ahead is winding. We are at a moment of real darkness, but there is no turning back. We may kick against this reality, but actually, it is a moment of truth, just as we find in all good stories. What is needed now is a steadfast commitment to the task and an understanding that failure is not an option.

We can be informed by art, literature and history as much as by science. Meeting the challenge of climate change needs to become part of a new story of human striving and renewal.

Right now, the predominant stories we are telling ourselves about the climate crisis are not very inspiring. But a new story can reinvigorate our efforts.

When the story changes, everything changes.

In October 1957, Americans looked upwards as the Soviet Union's *Sputnik* satellite crossed over the country.[1] For the first time, there was a satellite in the sky, and their 'enemy' had beaten them to it. That night, from Pennsylvania to Kansas to Colorado, families realised in dismay that the enemy could see them, was watching them.

How did the country respond? Within a few years, President John F. Kennedy gave his famous speech about landing a man on the moon within that decade, a feat far more challenging than launching a satellite.[2] He spoke of it without knowing whether it could be done, and without a detailed budget, plan or timeline. He was reclaiming the narrative and placing Americans inside a story that was hopeful and in which they could prevail.

The speech both terrified and electrified NASA. Within a few months it reorganised itself in line with this new goal. Teams worked harder than ever to innovate, which was particularly galvanising and thrilling for young people; the average age of the team that launched the Apollo missions was 28.[3] Everyone was part of a shared endeavour that gave their lives meaning.

When Kennedy first paid a visit to NASA Mission Control, at one point he came across a janitor who was cleaning the control room. 'And what is your role here?' he asked.

'Mr President, sir,' came the reply, 'I'm putting a man on the moon.'[4]

The compelling vision made this man feel that he was part of something great, and he was. Someone had to keep the room clean: it would not have been possible to put a man on the moon if that didn't happen. Imagine how the janitor would have felt, however, if he had been cleaning a control room for a government agency that had been bested by a rival and was facing relative decline. It was the story that motivated him to action.

Consider also the story that Great Britain told itself as it was enduring the blitzkrieg raids of 1941. As late as 1939, Britain had torn itself to pieces over different ideas of how to deal with Hitler. Prime Minister Neville Chamberlain was committed to a policy of appeasement and had great support. With the memories of the First World War still fresh, a good proportion of people would have done anything to avoid facing the reality that Hitler would stop at nothing to conquer Europe. Eventually, Chamberlain fell, and in his place came Winston Churchill. Churchill is remembered for many things, not all of them positive, but his most remarkable achievement in those early days was embedding a new story into the national psyche that prepared people for what was to come. An island alone. A greatest hour. A greatest generation that would fight them on the beaches and fight them in the hills and in the streets. A country that would never surrender.

Countless interviews with those who lived through that time have again and again described how a spirit of shared endeavour infused all actions, from the pilots in the Battle of Britain, to the people who turned their gardens and green spaces into food production on a massive scale. The simple task of pulling potatoes from the soil became an act of service in support of absent loved ones at the front and part of the pursuit of victory.

Even with the Paris Agreement, for the longest time, the story that prevailed was that climate change was too complicated; it was impossible to get countries to agree, and the structure of the UN would not allow agreement. The negotiations were populated with thousands of people who could explain in great detail and for many hours why there was absolutely no way through the myriad complexity to reach agreement. Changing that mindset was the hardest but most critical step we took. The journey from the failure in Copenhagen to the culmination in Paris was marked by a gradual build-up of momentum, and as the momentum built, the story changed.

At first there were only a few, but over time, thousands of people became convinced that the moment for progress was possible and that they had an important role to play. As each country made a commitment, more people believed in this possibility. The price of solar panels fell, cities took leadership positions, people marched in the streets, corporations took action and investors moved money out of fossil fuels. They all became steps on the journey to a new story.

At this moment, when we have reached the limits of the planet's ability to sustain life in the form in which we know it, we have also reached the limits of the stories that define our lives. Personal achievements through individualistic competition, continuous consumption, scepticism about our ability to come together as humanity and an inability to see the deeper impacts of what we are doing to the planet – all are no longer useful.

Now we must move towards understanding our shared existence on this planet, not because it is a nice addendum to what we do but because it is a matter of survival. Our current quest for a regenerative future has even higher levels of complexity and is decisively more consequential than the US quest to put a man on the moon or the UK's determination to defeat Hitler.

This is not the quest of one nation. This time it's up to all of us, to all the nations and peoples of the world. No matter how complex or deep our differences, we fundamentally share everything that is important: the desire to forge a better world for everyone alive today and all the generations to come.

Imagine, just for a moment, a world in which we had achieved this quest. It may seem far-fetched to you, utopian even, but since the very survival of humanity is at stake, ironically we believe that our chances of rising to this challenge are greater now than they have ever been. Humanity is capable of coming together to do this. Whether we will succeed in doing so will become apparent in a few short years.

With this book, we have begun to weave together some of the elements of our new story.

We can, together, reimagine our place in this world. As human beings, we all have the outrageous fortune to be here on this planet at this moment of profound consequence.

When the eyes of our children, and their children, look straight into ours, and they ask us 'What did you do?' our answer cannot just be that we did everything we could.

It has to be more than that.

There is really only one answer.

We did everything that was necessary.

So let us begin today to tell the story of how we did not baulk at this seemingly insurmountable challenge, of how we were not defeated by the multiple setbacks we encountered. Let us tell the story of how we made the choice to pull away from the brink of peril, of how we took our responsibility seriously and did everything that was necessary to emerge from the crisis while rekindling our relationships with each other and with all the natural systems that enable human life on Earth.

Let it be a story of great adventure, against overwhelming odds.

A story of survival.

And of a thriving existence.

What You Can Do Now

This action plan is part of a growing movement of Stubborn Optimists committed to fulfilling the vision of a regenerative world. We can only do this together and hope you will join us at www.GlobalOptimism.com

Right Now

|| Take a deep breath and decide that collectively we can do this, and that you will play your part. You will be a hopeful visionary for humanity through these dark days. From this moment, despair ends and tactics begin.

|| Decide that you will be part of the politics of the future. You will vote for, campaign for, and support candidates who champion emissions reductions. Reject the politics of nostalgia. For the next ten years, this will be your number-one political priority.

|| Commit to reducing your impact on the climate by more than half of what it is today by 2030. Aim for 60 per cent.

Just because right now you don't know how you will do so does not need to stop you. We are all learning.

Today or Tomorrow

|| Determine where your principal elected officials stand on climate change; write to them about your commitments, and let them know. Tell them you are watching.

|| Choose at least one day of the week to go meat-free, and decide how soon you will add more days to that commitment.

|| Think big. How do you most impact climate change, and what big things can you do to effect a regenerative future?

|| Tell others about your commitments, in person or on social media. Don't be shy! Invite others to follow suit. Your example will motivate them.

This Week

|| Share your personal plan to reduce emissions by more than half with your partner, kids and friends, and invite them to do so as well. Preserving the future of all life should be joyful. Have fun with it.

|| Take some actions and stick to them over time – it will give you momentum. Reduce daily energy use, bike instead of driving a car, switch your energy supplier to 100 per cent clean. It's all good and all needs doing. Consider what else you can do, while remembering there is still much to be done.

‖ Go outside and look around. This world is damaged and hurting, but it is also beautiful and intact and whole. Pay attention to something you have forgotten – emerging leaves in the spring or frost on dead leaves in winter. Feel the gratitude we owe the Earth for her bounty and beauty.

‖ Tune into our podcast for a healthy dose of 'Outrage & Optimism' where we chat about all things climate with special guests and experts from all over the world. Outrage and Optimism is available wherever you get your podcasts, or via https://globaloptimism.com/podcasts/

This Month

‖ Find out who in your vicinity is organising political action involving climate change. Attend meetings and meet the concerned citizens. Go to demonstrations and marches! Allow yourself to be inspired by the miracle of committed groups intent on changing the world.

‖ Start a conversation with someone who is not active on climate change with a view towards understanding their stance and gently enlarging their awareness of the crisis from their perspective.

‖ Enact your commitments: What precisely will you do this year? How will it affect you and your family? How will you begin to apply the changes you plan to make?

‖ Calculate your carbon footprint so you can see where your emissions come from. There are several different tools available online, and we've listed a couple in the following section 'Where to Go Next'. Choose one that suits you and use it to understand exactly where you can take action for the biggest impact.

|| Register your commitments at www.count-us-in.org, and join a growing community of people from around the world who are counting down their emissions together.

|| Challenge your consumerism. Look at what you have bought, and ask yourself whether it brings you joy. Question your impulses to buy more, and begin to see how liberating it is to buy less.

|| Start a mindfulness practice, perhaps a breathing exercise of gratitude. Do it every day, if only for a few minutes. Learn to create a gap of light between yourself, the world and your reactions.

|| Plant trees. As many as you can. Look for a local group doing tree planting. Get out there when you can, and when you can't, support others to do so.

|| Understand your privilege in relation to others, and commit to helping level the playing field for all.

This Year

|| Be political in your daily life. Seek collective opportunities to advance the cause of emissions reductions. It will inspire you and help you feel you are part of a shared endeavour. Engage regularly in direct action if that is possible where you live. VOTE!

|| Be consistent. You may have changed your electricity supply to 100 per cent renewable energy, rethought your commute, changed your air travel habits and altered your diet. If you can sustain your effort for the first year, you stand a good chance of doing so every year. Recognise your accomplishment.

By 2030

|| Deliver on your plan to cut your emissions by more than half. Celebrate your achievement.

|| Finance others to plant more trees as a symbol of the fact that you still have some way to go. Trees are good, and the world needs more of them.

|| Ensure you have voted in line with these priorities in national and regional elections and been vocal about the fact that you have done so.

|| Continue to practise the other new habits you have developed.

|| Encourage those closest to you – family, friends, loved ones – to be climate conscious.

|| Start the plan to reduce your emissions again by more than half over the next decade.

Before 2050

|| Be at net-zero emissions, having been part of the generation that chose a better future for all of us.

Where To Go Next

Activism

350.org
www.350.org

Break Free From Plastic
www.breakfreefromplastic.org

Extinction Rebellion
www.rebellion.earth

Greenpeace
www.greenpeace.org.uk/challenges/climate-change

Possible
www.wearepossible.org

People & Planet
www.peopleandplanet.org

UK Youth Climate Coalition
www.ukycc.com

Fridays For Future
www.fridaysforfuture.org

Advocacy

Citizens' Climate Lobby
www.citizensclimatelobby.org

Citizen Science
www.Citizenscience.org

Client Earth
www.Clientearth.org

Climate Action Network International
www.climatenetwork.org

The Climate Group
www.theclimategroup.org

Committee on Climate Change UK
www.theccc.org.uk

Friends of the Earth
www.friendsoftheearth.uk/climate-change

Future Stewards
www.leadersquest.org/future-stewards

Global Optimism
www.globaloptimism.com

Go Fossil Free
www.gofossilfree.org

Plastic Pollution Coalition
www.plasticpollutioncoalition.org

World Resources Institute
www.wri.org/our-work

Carbon calculators and other tools

Cool Climate Network
coolclimate.berkeley.edu/calculator

Ecosia – the search engine that plants trees
www.Ecosia.org

Foodprint
www.Foodprint.org

Global Footprint Network
www.footprintnetwork.org/resources/footprint-calculator

How to talk about climate change
www.weforum.org/agenda/2018/06/how-to-talk-about-climate-
    change-5-tips-from-the-front-lines

United Nations Carbon Offset Platform
www.offset.climateneutralnow.org

Nature and Conservation

Born Free
www.bornfree.org.uk

Earth Alliance
Ealliance.org

Dr Jane Goodall
www.janegoodall.org

Rewilding Europe
www.rewildingeurope.com

World Land trust
www.worldlandtrust.org/appeals/scorched-earth

WWF (World Wildlife Fund)
www.wwf.org.uk

The Science

Earth Observatory, NASA
www.earthobservatory.nasa.gov

National Geographic
nationalgeographic.com

Nature: Climate Change
nature.com

Our world in data
Ourworldindata.org

Project Drawdown
Drawdown.org

Science Alert
Sciencealert.com

Science Direct
Sciencedirect.com

Smithsonian Magazine
smithsonianmag.com

*Skeptical Science: Getting skeptical
about global warming scepticism*
www.skepticalscience.com

IPCC (The Intergovernmental Panel on Climate Change)
www.ipcc.ch

Water Scarcity Atlas
www.waterscarcityatlas.org

World Health Organisation
www.who.int

Tipping Points

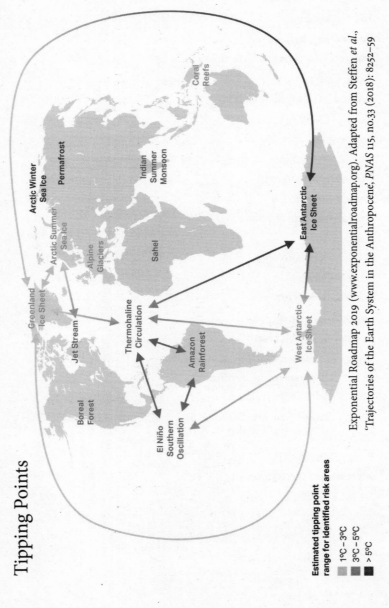

Estimated tipping point range for identified risk areas

1°C – 3°C
3°C – 5°C
> 5°C

Exponential Roadmap 2019 (www.exponentialroadmap.org). Adapted from Steffen *et al.*, 'Trajectories of the Earth System in the Anthropocene', *PNAS* 115, no.33 (2018): 8252–59

Temperature Scenarios

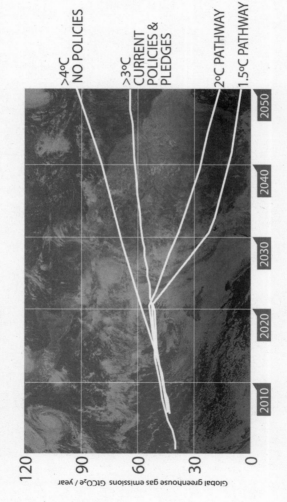

Adapted from Climate Action Tracker (https://climateactiontracker.org/global/temperatures/)

Acknowledgements

First, we would like to thank those family members and mentors who have shaped and guided our worldviews. Among them are José Figueres Ferrer, Kofi Annan, Thich Nhat Hanh, Bee Rivett-Carnac, Nigel Topping, Antony Turner, Paul Dickinson, Fraser Durham, Howard and Sue Lamb, Vivienne and Michael Zammit Cutajar, Sister True Dedication, Phap Lai, and Phap Linh.

This book is in many ways an outcome of the work of all those people who co-created the Paris Agreement of 2015, and of the many efforts since then to ensure we meet the challenge of our times.

A significant group of trusted friends and advisors helped us develop and hone the ideas in the book in a direct way. We are grateful to them all for their patient reviewing and wise counsel. In particular we would like to mention Natasha Rivett-Carnac, Jesse Abrams, Stephanie Antonian, Rosina Birbaum, Amanda Eichel, Nick Foster, Thomas Friedman, Sarah Goodenough, Callum Grieve, Dave Hicks, Andrew

Higham, John Holdren, Sarah Hunter, Merlin Hyman, Raj Joshi, Andy Karsner, Satish Kumar, Graham Leicester, Lindsay Levin, Thomas Lingard, Thomas Lovejoy, Mark Lynas, Michael Mann, Marina Mansilla Hermann, Mark Maslin, Bill McKibben, Jennifer Morgan, Jules Peck, Matthew Phillips, Brooks Preston, Shyla Raghav, Chloe Revill, Mike Rivett-Carnac, Bill Sharpe, Nicholas Stern, Betsy Taylor, Anne Topping, Patrick Verkooijen, Steve Waygood, Daniel Wahl, Martin Weinstein, and Kerem Yilmaz. Extra special thanks are due to Zoe Tcholak-Antitch, Lauren Hamlin, and Victoria Harris.

A much larger group of friends and colleagues have been our fellow travellers both in the creation of the Paris Agreement and in the vital next steps the world is now taking to address the climate crisis and deliberately choose a better future. This list is vast, and it would be impossible for us to mention everyone here, but we would like to pay special mention to Alejandro Agag, Lorena Aguilar, Fahad Al Attiya, Ali Al-Naimi, Carlos Alvarado Quesada, Ken Alex, Christiane Amanpour, Chris Anderson, Mats Andersson, Monica Araya, John Ashford, David Attenborough, AURORA, Mariana Awad, Peter Bakker, Vivian Balakrishnan, Ajay Banga, Greg Barker, Ecumenical Patriarch Bartholomew, Kevin Baumert, Sue Biniaz, Fatih Birol, Nicolette Bartlett, Oliver Bäte, Marc Benioff, Jeff Bezos, Dean Bialek, Michael Bloomberg, May Boeve, Gail Bradbrook, Piers Bradford, Richard Branson, Jesper Brodin, Tom Brookes, Jerry Brown, Sharan Burrow, Felipe Calderon, Kathy Calvin, Mark Campanale,

Miguel Arias Cañete, Mark Carney, Clay Carnill, Andrea Correa do Lago, Anne-Sophie Cerisola, Robin Chase, Sagarika Chatterjee, Pilita Clark, Helen Clarkson, Jo Confino, Aron Cramer, David Crane, Tomas Anker Christensen, John Danilovich, Conyers Davis, Tony de Brum, Bernaditas de Castro Muller, Brian Deese, Claudio Descalzi, Leonardo DiCaprio, Paula DiPerna, Elliot Diringer, Sandrine Dixson Decleve, Ahmed Djoghlaf, Claudia Dobles Camargo, Alister Doyle, José Manuel Entrecanales, Hernani Escobar, Patricia Espinosa, Emmanuel Faber, Nathan Fabian, Laurent Fabius, Emily Farnworth, Daniel Firger, James Fletcher, Pope Francis, Gail Gallie, Grace Gelder, Kristalina Georgieva, Cody Gildart, Jane Goodall, Al Gore, Kimo Goree, Ellie Goulding, Mats Granryd, Jerry Greenfield, Ólafur Grímsson, Sally Grover Bingham, Emmanuel Guerin, Kaveh Guilanpour, Stuart Gulliver, Angel Gurria, Antonio Guterres, William Hague, Thomas Hale, Brad Hall, Winnie Hallwachs, Simon Hampel, Kate Hampton, Yuval Noah Harari, Jacob Heatley-Adams, Julian Hector, Hilda Heine, Ned Helme, Barbara Hendricks, Jamie Henn, Anne Hidalgo, François Hollande, Emma Howard Boyd, Stephen Howard, Arianna Huffington, Kara Hurst, Jay Inslee, Natalie Isaacs, Maria Ivanova, Lisa Jackson, Lisa Jacobson, Dan Janzen, Michel Jarraud, Sharon Johnson, Kelsey Juliana, Yolanda Kakabadse, Lila Karbassi, Caio Koch-Weser, Marcin Korolec, Iain Keith, Mark Kenber, John Kerry, Sean Kidney, Jim Kim, Ban Ki-moon, Lise Kingo, Richard Kinley, Sister Jayanti Kirpalani, Isabelle Kocher, Larry Kramer, Kalee Kreider, Kishan Kumarsingh,

Rachel Kyte, Christine Lagarde, Philip Lambert, Dan Lashof, Guilherme Leal, Penelope Lea, Bernice Lee, Jeremy Leggett, Thomas Lingard, Andrew Liveris, Hunter Lovins, Mindy Lubber, Miguel Ángel Mancera Espinosa, Stella McCartney, Gina McCarthy, Bill McDonouh, Catherine McKenna, Sonia Medina, Bernadette Meehan, Johannes Meier, Maria Mendiluce, Antoine Michon, David Miliband, Ed Miliband, Amina Mohammed, Jennifer Morris, Tosi Mpanu-Mpanu, Nozipho Mxakato-Diseko, Kumi Naidoo, Nicole Ng, Maite Nkoana-Mashabane, Indra Nooyi, Michael Northrop, Tim Nuthall, Bill Nye, Rafe Offer, Jean Oelwang, Ngozi Okonjo-Iweala, Hindou Oumarou Ibrahim, Mo Ibrahim, Kevin O Hanlon, René Orellana, Ricken Patel, Jose Penido, Charlotte Pera, Jonathan Pershing, Stephen Petricone, Stephanie Pfeifer, Shannon Phillips, Bertrand Piccard, François-Henri Pinault, John Podesta, Paul Polman, Ian Ponce, Carl Pope, Jonathon Porritt, Patrick Pouyanne, Manuel Pulgar Vidal, Tracy Raczek, Jairam Ramesh, Curtis Ravenell, Robin Reck, Geeta Reddy, Dan Reifsnyder, Fiona Reynolds, Alex Rivett-Carnac, Chris Rivett-Carnac, Ben Rhodes, Nick Robins, Jim Robinson, Mary Robinson, Cristiam Rodriguez, Matthew Rodriguez, Kevin Rudd, Mark Ruffalo, Artur Runge-Metzger, Fredric Samama, Karsten Sach, Claudia Salerno Caldera, Richard Samans, M. Sanjayan, Steve Sawyer, Jerome Schmitt, Kirsty Schneeberger, Klaus Schwab, Arnold Schwarzernegger, Jeff Seabright, Maros Sefcovic, Leah Seligmann, Peter Seligmann, Oleg Shamanov, Kevin Sheekey, Seth Schultz, Feike Sijbesma, Nat Simons, Paul Simpson, Michael Skelly,

Erna Solberg, Andrew Steer, Achim Steiner, Todd Stern, Tom Steyer, Irene Suárez, Mustafa Suleyman, Terry Tamminen, Ratan Tata, Astro Teller, Tessa Tenant, Susan Tierney, Halldór Thorgeirsson, Greta Thunberg, Svante Thunberg, Halla Tomasdottir, Laurence Tubiana, Keith Tuffley, Jo Tyndall, Hamdi Ulukaya, Ben van Beurden, Andy Vesey, Gino van Begin, Mark Watts, Dominic Waughray, Meridith Webster, Scott Weiner, Helen Wildsmith, Antha Williams, Dessima Williams, Mark Wilson, Justin Winters, Martin Wolf, Farhana Yamin, Zhang Yue, Mohammed Yunus, Jochen Zeitz, and Xie Zhenhua.

We would like to thank each and every one of the outstanding colleagues of the secretariat of the United Nations Framework Convention on Climate Change, the always thorough UN security personnel, and the exemplary Mission 2020 team.

This book would not have been possible without the remarkable skills of the editors at Knopf and Bonnier that we were privileged to work with, Erroll McDonald and Margaret Stead, and their respective teams.

After spending a good two years thinking about writing a book and making almost no progress, the step up occurred when we met Doug Abrams in September 2018. Doug and the team at Idea Architects transformed our approach and made the project real in a way it simply would never have been without them. In many ways, the book owes its genesis to this team more than any other and, alongside Doug, to wordsmith Lara Love and efficient Ty Gideon Love.

ACKNOWLEDGEMENTS

Our gratitude goes also to Caspian Dennis, Sandy Violette and the whole team at Abner Stein, as well as Camilla Ferrier, Jemma McDonagh and the entire team at the Marsh Agency.

Finally, we cannot end this acknowledgement without thanking the close friends and family members who supported us through the writing of this book. The few months of actual writing time were marked by a remarkable intensity of major events in our lives, of both sadness and joy. These included the passing of two of Christiana's brothers, Mariano and Martí; of Tom's mother-in-law, Irene Walter; and of Doug's father, Richard Abrams. It also included the wedding of Christiana's daughter Yihana. We are left with a deep sense of gratitude towards those closest to us who generously and patiently supported us throughout this period, in particular Naima Ritter, Yihana Ritter, Kirsten Figueres, Mariano Figueres, Chaco Delgado, David Hall, Ron Walter, Diana Strike, Sara Rivett-Carnac and Natasha Rivett-Carnac.

You are our past, our present and our future.

Notes

Introduction

1   Charles D. Keeling, 'The Concentration and Isotopic
    Abundances of Carbon Dioxide in the Atmosphere',
    Tellus 12, no. 2 (May 1960): 200–3, https://doi.org/10.1111/
    j.2153-3490.1960.tb01300.x. The Scripps Institution of
    Oceanography at UC Davis has kept records of global
    atmospheric carbon dioxide concentration since 1958,
    updating the Keeling Curve: https://scripps.ucsd.edu/
    programs/keelingcurve/.

Chapter 1: Choosing Our Future

1   For more on ice ages, see, for example, Michael
    Marshall, 'The History of Ice on Earth', *New Scientist*,
    24 May 2010, https://www.newscientist.com/article/
    dn18949-the-history-of-ice-on-earth/.

2   The world's population is expected to hit 9.8 billion by
    2050. United Nations Department of Economic and Social
    Affairs, 'Growing at a Slower Pace, World Population Is

Expected to Reach 9.7 Billion in 2050 and Could Peak at
Nearly 11 Billion around 2100', 17 June 2019, https://www.
un.org/development/desa/en/news/population/world-
population-prospects-2019.html.

3   Daniel Christian Wahl, 'Learning from Nature
and Designing as Nature: Regenerative Cultures
Create Conditions Conducive to Life', Biomimicry
Institute, 6 September 2016, https://biomimicry.org/
learning-nature-designing-nature-regenerative-cultures-
create-conditions-conducive-life/.

4   The Industrial Revolution and the explosion of fossil
fuel use changed our direction. For more on this, see
History.com, 'Industrial Revolution', 1 July 2019 (updated 9
September 2019), https://www.history.com/topics/industrial-
revolution/industrial-revolution for a history of the
Industrial Revolution; and Hannah Ritchie and Max Roser,
'Fossil Fuels', Our World in Data, https://ourworldindata.
org/fossil-fuels, for the development of fossil fuel use.

5   National Aeronautics and Space Administration, 'Changes
in the Carbon Cycle', NASA Earth Observatory, 16 June 2011,
https://earthobservatory.nasa.gov/features/CarbonCycle/
page4.php.

6   Rémi d'Annunzio, Marieke Sandker, Yelena Finegold and
Zhang Min, 'Projecting Global Forest Area towards 2030',
Forest Ecology and Management 352 (2015): 124–33, https://
www.sciencedirect.com/science/article/pii/S0378112715001346;
John Vidal, 'We Are Destroying Rainforests So Quickly
They May Be Gone in 100 Years', *Guardian* (US edition),

23 January 2017, https://www.theguardian.com/global-
development-professionals-network/2017/jan/23/
destroying-rainforests-quickly-gone-100-years-deforestation.

7   Josh Gabbatiss, 'Earth Will Take Millions of Years to Recover
from Climate Change Mass Extinction, Study Suggests',
Independent, 8 April 2019, https://www.independent.co.uk/
environment/mass-extinction-recovery-earth-climate-change-
biodiversity-loss-evolution-a8860326.html.

8   Richard Grey, 'Sixth Mass Extinction Could Destroy Life as
We Know It – Biodiversity Expert', *Horizon*, 4 March 2019,
https://horizon-magazine.eu/article/sixth-mass-extinction-
could-destroy-life-we-know-it-biodiversity-expert.html;
Gabbatiss, 'Earth Will Take Millions of Years.'

9   LuAnn Dahlman and Rebecca Lindsey, 'Climate Change:
Ocean Heat Content', Climate.gov, 1 August 2018, https://
www.climate.gov/news-features/understanding-climate/
climate-change-ocean-heat-content.

10 Lauren E. James, 'Half of the Great Barrier Reef Is
Dead', *National Geographic*, August 2018, https://www.
nationalgeographic.com/magazine/2018/08/explore-atlas-
great-barrier-reef-coral-bleaching-map-climate-change/.

11 T. Schoolmeester, H. L. Gjerdi, J. Crump, *et al.*, *Global
Linkages: A Graphic Look at the Changing Arctic (rev.
1)* (Nairobi and Arendal: UN Environment and GRID-
Arendal, 2019), http://www.grida.no/publications/431.

12 National Aeronautics and Space Administration, 'As Seas
Rise, NASA Zeros In: How Much? How Fast?' 3 August 2017,
https://www.nasa.gov/goddard/risingseas.

13 Joseph Stromberg, 'What Is the Anthropocene and Are We in It?' *Smithsonian*, January 2013, https://www.smithsonianmag.com/science-nature/what-is-the-anthropocene-and-are-we-in-it-164801414/.

14 An exploration can be found in Darrell Moellendorf, 'Progress, Destruction, and the Anthropocene', *Social Philosophy and Policy* 34, no. 2 (2017): 66–88. See also the documentary film *Anthropocene: The Human Epoch*, 2018, https://theanthropocene.org/film/.

15 More than 3 degrees Celsius warmer than pre-industrial average global temperature.

16 That is, 1.5 degrees Celsius higher than pre-industrial average global temperature.

17 For a full explanation, see Intergovernmental Panel of Climate Change, 'Special Report: Global Warming of 1.5 °C', 2018, https://www.ipcc.ch/sr15/.

18 Nebojsa Nakicenovic and Rob Swart, *eds*, *Special Report on Emissions Scenarios* (Cambridge: Cambridge University Press, 2000), https://www.ipcc.ch/report/emissions-scenarios/.

Chapter 2: The World We Are Creating

1 Department of Public Health, Environmental and Social Determinants of Health, World Health Organisation, 'Ambient Air Pollution: Health Impacts', https://www.who.int/airpollution/ambient/health-impacts/en/.

2 Greenpeace Southeast Asia, 'Latest Air Pollution Data Ranks World's Cities Worst to Best', 5 March 2019,

https://www.greenpeace.org/southeastasia/press/679/
latest-air-pollution-data-ranks-worlds-cities-worst-to-best/.

3 'Cloud Seeding', ScienceDirect, https://www.sciencedirect.com/
topics/earth-and-planetary-sciences/cloud-seeding.

4 Acid rain is any form of precipitation that contains high
levels of nitric and sulfuric acids. It can also occur in
the form of snow and fog. Normal rain is slightly acidic,
with a pH of 5.6, while acid rain has a pH between 4.2
and 4.4. Most acid rain is a product of human activities.
The biggest sources are coal power plants, factories and
vehicles. See Christina Nunez, 'Acid Rain Explained',
National Geographic, 28 February 2019, https://www.
nationalgeographic.com/environment/global-warming/
acid-rain/.

5 Heather Smith, 'Will Climate Change Move Agriculture
Indoors? And Will That Be a Good Thing?' Grist, 3
February 2016, https://grist.org/food/will-climate-change-
move-agriculture-indoors-and-will-that-be-a-good-thing/.

6 Johan Rockström, 'Climate Tipping Points', Global
Challenges Foundation, https://www.globalchallenges.org/
en/our-work/annual-report/climate-tipping-points.

7 See David Wallace-Wells, *The Uninhabitable Earth: Life
after Warming* (New York: Tim Duggen Books, 2019).

8 Great Barrier Reef Marine Park Authority, 'Climate
Change', 2018, http://www.gbrmpa.gov.au/our-work/
threats-to-the-reef/climate-change.

9 Aylin Woodward, 'One of Antarctica's Biggest Glaciers
Will Soon Reach a Point of Irreversible Melting', *Business*

Insider France, 9 July 2019, http://www.businessinsider.fr/
us/antarctic-glacier-on-way-to-irreversible-melt-2019-7.

10   Roz Pidcock, 'Interactive: What Will 2C and 4C of
Warming Mean for Sea Level Rise?' Carbon Brief,
11 September 2015, https://www.carbonbrief.org/
interactive-what-will-2c-and-4c-of-warming-mean-for-
global-sea-level-rise; Josh Holder, Niko Kommenda,
and Jonathan Watts, 'The Three-Degree World: The
Cities That Will Be Drowned by Global Warming',
Guardian (US edition), 3 November 2017, https://www.
theguardian.com/cities/ng-interactive/2017/nov/03/
three-degree-world-cities-drowned-global-warming.

11   United Nations Climate Change News, 'Climate Change
Threatens National Security, Says Pentagon', 14 October
2014, https://unfccc.int/news/climate-change-threatens-
national-security-says-pentagon. For more useful resources,
see American Security Project, 'Climate Security Is
National Security', https://www.americansecurityproject.org/
climate-security/.

12   Polar Science Centre, 'Antarctic Melting
Irreversible in 60 Years', http://psc.apl.uw.edu/
antarctic-melting-irreversible-in-60-years/.

13   Ocean Portal Team, 'Ocean Acidification', Smithsonian
Institute, April 2018, https://ocean.si.edu/ocean-life/
invertebrates/ocean-acidification.

14   Chang-Eui Park, Su-Jong Jeong, Manoj Joshi, *et al.*,
'Keeping Global Warming Within 1.5 °C Constrains

Emergence of Aridification', *Nature Climate Change* 8, no. 1 (January 2018): 70–74.

15   Regan Early, 'Which Species Will Survive Climate Change?' *Scientific American*, 17 February 2016, https://www.scientificamerican.com/article/which-species-will-survive-climate-change/.

16   Scientific Expert Group on Climate Change and Sustainable Development, 'Confronting Climate Change: Avoiding the Unmanageable and Managing the Unavoidable', Sigma Xi, February 2007, https://www.sigmaxi.org/docs/default-source/Programs-Documents/Critical-Issues-in-Science/executive-summary-of-confronting-climate-change.pdf.

17   For more on the risks of climate change on these river systems, see John Schwartz, 'Amid 19-Year Drought, States Sign Deal to Conserve Colorado River Water', *New York Times*, 19 March 2019, https://www.nytimes.com/2019/03/19/climate/colorado-river-water.html; Sarah Zielinski, 'The Colorado River Runs Dry', *Smithsonian*, October 2010, https://www.smithsonianmag.com/science-nature/the-colorado-river-runs-dry-61427169/; 'Earth Matters: Climate Change Threatening to Dry Up the Rio Grande River, a Vital Water Supply', CBS News, 22 April 2019, https://www.cbsnews.com/news/earth-day-2019-climate-change-threatening-to-dry-up-rio-grande-river-vital-water-supply/.

18   Gary Borders, 'Climate Change on the Rio Grande', *World Wildlife Magazine*, Fall 2015, https://www.

worldwildlife.org/magazine/issues/fall-2015/articles/
climate-change-on-the-rio-grande.

19 Brian Resnick, 'Melting Permafrost in the Arctic Is
Unlocking Diseases and Warping the Landscape', Vox, 26
September 2019, https://www.vox.com/2017/9/6/16062174/
permafrost-melting.

20 'How Climate Change Can Fuel Wars', *Economist*, 23 May
2019, https://www.economist.com/international/2019/05/23/
how-climate-change-can-fuel-wars.

21 Silja Klepp, 'Climate Change and Migration',
Oxford Research Encyclopedias: Climate Science,
April 2017, https://oxfordre.com/climatescience/
view/10.1093/acrefore/9780190228620.001.0001/
acrefore-9780190228620-e-42.

22 Brian Resnick, 'Melting Permafrost in the Arctic Is Unlocking
Diseases and Warping the Landscape', Vox, 26 September 2019,
https://www.vox.com/2017/9/6/16062174/permafrost-melting.

23 Derek R. MacFadden, Sarah F. McGough, David
Fisman, Mauricio Santillana, and John S. Brownstein,
'Antibiotic Resistance Increases with Local Temperature',
Nature, 21 May 2018, https://www.nature.com/articles/
s41558-018-0161-6.

Chapter 3: The World We Must Create

1 P. J. Marshall, 'Reforestation: The Critical Solution
to Climate Change', Leonardo DiCaprio Foundation,
7 December 2018, https://www.leonardodicaprio.org/
reforestation-the-critical-solution-to-climate-change/.

2   Julio Díaz, public health and environment expert at the National School of Public Health, which is part of the Carlos III Health Institute, reports that individuals with kidney problems and neurodegenerative diseases, such as Parkinson's, visit the doctor more frequently in hot weather. Excessive heat also increases the risk of premature births and low birth rates. Cited in Manuel Planelles, 'More Than a Feeling: Summers in Spain Really Are Getting Longer and Hotter', *El País*, 3 April 2019, https://elpais.com/elpais/2019/04/03/inenglish/1554279672_888064.html.

3   E. O. Wilson Biodiversity Foundation, 'Half-Earth: Our Planet's Fight for Life', https://eowilsonfoundation.org/half-earth-our-planet-s-fight-for-life/; Emily E. Adams, 'World Forest Area Still on the Decline', Earth Policy Institute, 31 August 2012, http://www.earth-policy.org/indicators/C56/forests_2012.

4   Project Drawdown, 'Tree Intercropping', https://www.drawdown.org/solutions/food/tree-intercropping; Project Drawdown, 'Silvopasture', https://www.drawdown.org/solutions/food/silvopasture.

5   Petra Todorovich and Yoav Hagler, 'High-Speed Rail in America', America2050, January 2011, http://www.america2050.org/pdf/HSR-in-America-Complete.pdf; Anton Babadjanov, 'Can We Replace Cross-Country Air with Rail Travel? Yes, We Can!' Seattle Transit Blog, 15 February 2019, https://seattletransitblog.com/2019/02/15/can-we-replace-cross-country-air-with-rail-travel-yes-we-can/.

6   Project Drawdown, 'Nuclear', https://www.drawdown.org/solutions/electricity-generation/nuclear. See also

Union of Concerned Scientists, 'Nuclear Power & Global Warming', 22 May 2015 (updated 8 November 2018), https://www.ucsusa.org/nuclear-power/ nuclear-power-and-global-warming.

7   RMIT University, 'Solar Paint Offers Endless Energy from Water Vapor', ScienceDaily, 14 June 2017, https://www. sciencedaily.com/releases/2017/06/170614091833.htm.

8   Global Water Scarcity Atlas, 'Desalination Powered by Renewable Energy', https://waterscarcityatlas.org/ desalination-powered-by-renewable-energy/.

9   Project Drawdown, 'Pasture Cropping', https://www. drawdown.org/solutions/coming-attractions/pasture-cropping. See also Taylor Mooney, 'What Is Regenerative Farming? Experts Say It Can Combat Climate Change', CBS News, 28 July 2019, https://www.cbsnews.com/news/ what-is-regenerative-farming-cbsn-originals/.

10 For more on climate change and food prices, see Nitin Sethi, 'Climate Change Could Cause 29% Spike in Cereal Prices: Leaked UN Report', *Business Standard*, 15 July 2019, https://www.business-standard.com/article/ current-affairs/climate-change-could-cause-29-spike-in-cereal-prices-leaked-un-report-119071500637_1.html.

11 For more on this concept, see Anna Behrend, 'What Is the True Cost of Food?' *Spiegel Online*, 2 April 2016, https://www.spiegel.de/international/tomorrow/ the-true-price-of-foodstuffs-a-1085086.html; Megan Perry, 'The Real Cost of Food', Sustainable Food Trust,

November 2015, https://sustainablefoodtrust.org/articles/
the-real-cost-of-food/.

12 Sarah Gibbens, 'Eating Meat Has "Dire" Consequences
    for the Planet, Says Report', *National Geographic*, 16
    January 2019, https://www.nationalgeographic.com/
    environment/2019/01/commission-report-great-food-
    transformation-plant-diet-climate-change/.

13 Fisheries and Aquaculture Department, Food and
    Agriculture Organisation of the United Nations, 'Climate
    Change Mitigation Strategies', 28 September 2016, http://
    www.fao.org/fishery/topic/166280/en.

14 Jennifer L. Pomeranz, Parke Wilde, Yue Huang,
    Renata Micha, and Dariush Mozaffarian, 'Legal and
    Administrative Feasibility of a Federal Junk Food and
    Sugar-Sweetened Beverage Tax to Improve Diet', *American
    Journal of Public Health*, 10 January 2018, https://ajph.
    aphapublications.org/doi/10.2105/AJPH.2017.304159; Arlene
    Weintraub, 'Should We Tax Junk Foods to Curb Obesity?'
    Forbes, 10 January 2018, https://www.forbes.com/sites/
    arleneweintraub/2018/01/10/should-we-tax-junk-foods-to-
    curb-obesity/; Mexico and Hungary are already piloting
    the idea of taxing junk food; see Julia Belluz, 'Mexico
    and Hungary Tried Junk Food Taxes – and They Seem
    to Be Working', Vox, 17 January 2018 (updated 6 April 2018),
    https://www.vox.com/2018/1/17/16870014/junk-food-tax.

15 This is already happening: 'China's Hainan Province to
    End Fossil Fuel Car Sales in 2030', Phys.org, 6 March 2019,

https://phys.org/news/2019-03-china-hainan-province-fossil-fuel.html.

16 This is already happening in the UK: Tom Edwards, 'ULEZ: The Most Radical Plan You've Never Heard Of', BBC News, 26 March 2019, https://www.bbc.com/news/uk-england-london-47638862.

17 Smart Energy International, 'Storage Advancements Fast-Track New Power Projects, Experts Say', 21 June 2018, https://www.smart-energy.com/news/energy-storage-new-power-projects/.

18 Adela Spulber and Brett Smith, 'Are We Building the Electric Vehicle Charging Infrastructure We Need?' *IndustryWeek,* 21 November 2018, https://www.industryweek.com/technology-and-iiot/are-we-building-electric-vehicle-charging-infrastructure-we-need.

19 Echo Huang, 'By 2038, the World Will Buy More Passenger Electric Vehicles Than Fossil-Fuel Cars', Quartz, 15 May 2019, https://qz.com/1618775/by-2038-sales-of-electric-cars-to-overtake-fossil-fuel-ones/; Jesper Berggreen, 'The Dream Is Over – Europe Is Waking Up to a World of Electric Cars', CleanTechnica, 17 February 2019, https://cleantechnica.com/2019/02/17/the-dream-is-over-europe-is-waking-up-to-a-world-of-electric-cars/.

20 We can already achieve this acceleration in 2019. See James Gilboy, 'The Porsche Taycan Will Do Zero-to-60 in 3.5 Seconds', The Drive, 17 August 2018, https://www.thedrive.com/news/22984/

the-porsche-taycan-will-do-zero-to-60-in-3-5-seconds;
and classic car retrofits are already starting to take
off: Robert C. Yeager, 'Vintage Cars with Electric-
Heart Transplants', *New York Times*, 10 January 2019,
https://www.nytimes.com/2019/01/10/business/electric-
conversions-classic-cars.html.

21 United Nations Department of Economic and Social Affairs,
'68% of the World Population Projected to Live in Urban
Areas by 2050, Says UN', 16 May 2018, https://www.un.org/
development/desa/en/news/population/2018-revision-of-
world-urbanisation-prospects.html.

22 David Dudley, 'The Guy from Lyft Is Coming for
Your Car', CityLab, 19 September 2016, https://
www.citylab.com/transportation/2016/09/
the-guy-from-lyft-is-coming-for-your-car/500600/.

23 Annie Rosenthal, 'How 3D Printing Could
Revolutionize the Future of Development', Medium,
1 May 2018, https://medium.com/@plus_socialgood/
how-3d-printing-could-revolutionize-the-future-of-
development-54a270d6186d; Elizabeth Royte, 'What
Lies Ahead for 3-D Printing?' *Smithsonian*, May 2013,
https://www.smithsonianmag.com/science-nature/
what-lies-ahead-for-3-d-printing-37498558/.

24 Marissa Peretz, 'The Father of Drones' Newest
Baby Is a Flying Car', *Forbes*, 24 July 2019, https://
www.forbes.com/sites/marissaperetz/2019/07/24/
the-father-of-drones-newest-baby-is-a-flying-car/.

25 The 'slow-cation' was already popular from the
seventeenth to the nineteenth centuries, in the form
of the 'Grand Tour.' Richard Franks, 'What Was the
Grand Tour and Where Did People Go?' Culture Trip, 4
December 2017, https://theculturetrip.com/europe/articles/
what-was-the-grand-tour-and-where-did-people-go/.

26 International Organisation for Migration mission
statement, https://www.iom.int/migration-and-
climate-change-0. See also Erik Solheim and William
Lacy Swing, 'Migration and Climate Change Need
to Be Tackled Together', United Nations Framework
Convention on Climate Change, 7 September 2018,
https://unfccc.int/news/migration-and-climate-change-
need-to-be-tackled-together.

27 Richard B. Rood, 'What Would Happen to the Climate
If We Stopped Emitting Greenhouse Gases Today?' The
Conversation, 11 December 2014. http://theconversation.
com/what-would-happen-to-the-climate-if-we-stopped-
emitting-greenhouse-gases-today-35011.

28 The 3D-printed version is already building houses at speed.
See Adele Peters, 'This House Can Be 3D-Printed for $4,000',
Fast Company, 12 March 2018, https://www.fastcompany.
com/40538464/this-house-can-be-3d-printed-for-4000.

Chapter 4: Who We Choose to Be

1 Joanna Macy and Chris Johnstone, *Active Hope: How to Face
the Mess We're in Without Going Crazy* (San Francisco: New
World Library, 2012), 32.

Chapter 5: Stubborn Optimism

1   Kendra Cherry, 'Learned Optimism', Verywell Mind, 25 July 2019, https://www.verywellmind.com/learned-optimism-4174101.

2   Jeremy Hodges, 'Clean Energy Becomes Dominant Power Source in UK', *Bloomberg*, 20 June 2019, https://www.bloomberg.com/news/articles/2019-06-20/clean-energy-is-seen-as-dominant-source-in-u-k-for-first-time.

3   Jordan Davidson, 'Costa Rica Powered by Nearly 100% Renewable Energy', EcoWatch, 6 August 2019, https://www.ecowatch.com/costa-rica-net-zero-carbon-emissions-2639681381.html.

4   Sammy Roth, 'California Set a Goal of 100% Clean Energy, and Now Other States May Follow Its Lead', *Los Angeles Times*, 10 January 2019, https://www.latimes.com/business/la-fi-100-percent-clean-energy-20190110-story.html.

5   Václav Havel, *Disturbing the Peace: A Conversation with Karel Huizdala* (New York: Vintage Books, 1991), 181–82.

6   Rebecca Solnit, *Hope in the Dark: Untold Histories, Wild Possibilities* (Chicago, Ill.: Haymarket Books, 2016), 4.

Chapter 6: Endless Abundance

1   Brad Lancaster, 'Planting the Rain to Grow Abundance', lecture at TEDxTucson, 6 March 2017, https://www.youtube.com/watch?v=I2xDZlpInik.

2   American Sociological Association, 'In Disasters, Panic Is Rare; Altruism Dominates', ScienceDaily,

8 August 2002, https://www.sciencedaily.com/
releases/2002/08/020808075321.htm.

3   Therese J. Borchard, 'How Giving Makes Us Happy',
Psych Central, 8 July 2018, https://psychcentral.com/blog/
how-giving-makes-us-happy/.

4   Wikipedia, 'November 2015 Paris Attacks', https://
en.wikipedia.org/wiki/November_2015_Paris_attacks.

Chapter 7: Radical Regeneration

1   Richard Louv, *Last Child in the Woods: Saving Our
Children from Nature-Deficit Disorder* (New York:
Algonquin, 2005).

2   Gregory Bateson, *Steps to an Ecology of Mind* (Chicago:
University of Chicago Press, 1972).

3   Daniel Christian Wahl, *Designing Regenerative Cultures*
(Dorset, UK: Triarchy Press, 2016), 267.

Chapter 8: Doing What Is Necessary

1   Even if we did the world would not stop warming. See Ute
Kehse, 'Global Warming Doesn't Stop When the Emissions
Stop', Phys.org, 3 October 2017, https://phys.org/news/2017-
10-global-doesnt-emissions.html.

2   Caitlin E. Werrell and Francesco Femia, 'Climate
Change Raises Conflict Concerns', *UNESCO Courier*,
2018, no. 2, https://en.unesco.org/courier/2018-2/
climate-change-raises-conflict-concerns.

3   'Germany on Course to Accept One Million Refugees
in 2015', *Guardian* (US edition), 7 December 2015,

https://www.theguardian.com/world/2015/dec/08/
germany-on-course-to-accept-one-million-refugees-in-
2015.

4 Benedikt Peters, '5 Reasons for the Far Right
Rising in Germany', *Süddeutsche Zeitung*,
https://projekte.sueddeutsche.de/artikel/politik/
afd-5-reasons-for-the-far-right-rising-in-germany-e403522/.

5 Project Drawdown is a great additional resource, and
outlines 100 solutions to reverse global warming.

6 Reality Check team, 'Reality Check: Which Form of
Renewable Energy Is Cheapest?' BBC News, 26 October
2018, https://www.bbc.com/news/business-45881551.

7 Michael Savage, 'End Onshore Windfarm Ban, Tories
Urge', *Guardian* (US edition), 30 June 2019, https://
www.theguardian.com/environment/2019/jun/30/
tories-urge-lifting-off-onshore-windfarm-ban.

8 Shannon Hall, 'Exxon Knew about Climate Change
Almost 40 Years Ago', *Scientific American*, 26 October
2015, https://www.scientificamerican.com/article/
exxon-knew-about-climate-change-almost-40-years-ago/.

9 Sarah Pruitt, 'How the Treaty of Versailles and German
Guilt Led to World War II', History.com, 29 June 2018
(updated 3 June 2019), https://www.history.com/news/
treaty-of-versailles-world-war-ii-german-guilt-effects.

10 S.P., 'What, and Who, Are France's "Gilets Jaunes"?'
Economist, 27 November 2018, https://www.
economist.com/the-economist-explains/2018/11/27/
what-and-who-are-frances-gilets-jaunes.

11   Alex Birkett, 'Online Manipulation: All the Ways You're
     Currently Being Deceived', Conversion XL, 19 November
     2015 (updated 7 February 2019), https://conversionxl.com/
     blog/online-manipulation-all-the-ways-youre-currently-
     being-deceived/.

12   Stephanie Pappas, 'Shrinking Glaciers Point to Looming
     Water Shortages', Live Science, 8 December 2011, https://
     www.livescience.com/17379-shrinking-glaciers-water-
     shortages.html.

13   Bridget Alex, 'Artic [sic] Meltdown: We're Already Feeling
     the Consequences of Thawing Permafrost', Discover, June
     2018, http://discovermagazine.com/2018/jun/something-stirs.

14   Fern Riddell, 'Suffragettes, Violence and Militancy',
     British Library, 6 February 2018, https://
     www.bl.uk/votes-for-women/articles/
     suffragettes-violence-and-militancy.

15   Office of the Historian, Department of State, 'The
     Collapse of the Soviet Union', https://history.state.gov/
     milestones/1989-1992/collapse-soviet-union.

16   'Futurama: "Magic City of Progress"' in World's
     Fair: Enter the World of Tomorrow, Biblion. http://
     exhibitions.nypl.org/biblion/worldsfair/enter-
     world-tomorrow-futurama-and-beyond/story/
     story-gmfuturama.

17   Abby Norman, 'Aliens, Autonomous Cars, and AI: This Is
     the World of 2118', Futurism.com, 11 January 2018, https://
     futurism.com/2118-century-predictions; Matthew Claudel
     and Carlo Ratti, 'Full Speed Ahead: How the Driverless

Car Could Transform Cities', McKinsey & Company, August 2015, https://www.mckinsey.com/business-functions/sustainability/our-insights/full-speed-ahead-how-the-driverless-car-could-transform-cities.

18 Brad Plumer, 'Cars Take Up Way Too Much Space in Cities. New Technology Could Change That', Vox, 2016, https://www.vox.com/a/new-economy-future/cars-cities-technologies; Vanessa Bates Ramirez, 'The Future of Cars Is Electric, Autonomous, and Shared – Here's How We'll Get There', Singularity Hub, 23 August 2018, https://singularityhub.com/2018/08/23/the-future-of-cars-is-electric-autonomous-and-shared-heres-how-well-get-there/.

19 Tim Walker, 'Maya Angelou Dies: "You May Encounter Many Defeats, but You Must Not Be Defeated"', Independent, 28 May 2014, https://www.independent.co.uk/news/people/maya-angelou-dies-you-may-encounter-many-defeats-but-you-must-not-be-defeated-9449234.html.

20 'Martin Luther King Jr. – Biography', NobelPrize.org, https://www.nobelprize.org/prizes/peace/1964/king/biographical.

21 Jonathan Swift, 'The Art of Political Lying', The Examiner, 9 November 1710. https://www.bartleby.com/209/633.html.

22 Soroush Vosoughi, Deb Roy and Sinan Aral, 'The Spread of True and False News Online', Science, 9 March 2018, https://science.sciencemag.org/content/359/6380/1146.full.

23 Carolyn Gregoire, 'The Psychology of Materialism, and Why It's Making You Unhappy', Huffington Post, 15 December 2013 (updated 7 December 2017), https://www.huffpost.com/entry/psychology-materialism_n_4425982.

24 Encyclopaedia Britannica Online, 'Confirmation Bias',
https://www.britannica.com/science/confirmation-bias.

25 Ben Webster, 'Britons Buy a Suitcase Full of New Clothes
Every Year', *Times* (UK), 5 October 2018, https://www.
thetimes.co.uk/article/britons-buy-a-suitcase-full-of-new-
clothes-every-year-wxws895qd.

26 United Nations Climate Change News, 'UN Helps
Fashion Industry Shift to Low Carbon', United
Nations Framework Convention on Climate
Change, 6 September 2018, https://unfccc.int/news/
un-helps-fashion-industry-shift-to-low-carbon.

27 Al Gore, *The Future: Six Drivers of Global Change* (New
York: Random House, 2013), 159.

28 Christina Gough, 'Super Bowl Average Costs of a 30-Second
TV Advertisement from 2002 to 2019 (in Million U.S.
Dollars)', Statista, 9 August 2019, https://www.statista.com/
statistics/217134/total-advertisement-revenue-of-super-bowls/.

29 Garett Sloane, 'Amazon Makes Major Leap in Ad Industry
with $10 Billion Year', AdAge, 31 January 2019, https://adage.
com/article/digital/amazon-makes-quick-work-ad-industry-
10-billion-year/316468.

30 A. Guttmann, 'Global Advertising Market – Statistics
& Facts', Statista, 24 July 2018, https://www.statista.com/
topics/990/global-advertising-market/.

31 A great article summing up the research can be found
here: Tori DeAngelis, 'Consumerism and Its Discontents',
American Psychological Association, June 2004, https://
www.apa.org/monitor/jun04/discontents.

32 Ibid.

33 Tony Seba and James Arbib, 'Are We Ready for the End of Individual Car Ownership?' *San Francisco Chronicle,* 10 July 2017, https://www.sfchronicle.com/opinion/openforum/article/Are-we-ready-for-the-end-of-individual-car-11278535.php.

34 A great article and podcast on this can be found here: Hans-Werner Kaas, Detlev Mohr, and Luke Collins, 'Self-Driving Cars and the Future of the Auto Sector', McKinsey & Company, August 2016, https://www.mckinsey.com/industries/automotive-and-assembly/our-insights/self-driving-cars-and-the-future-of-the-auto-sector.

35 Rosie McCall, 'Millions of Fossil Fuel Dollars Are Being Pumped into Anti-Climate Lobbying', IFLScience, 22 March 2019, https://www.iflscience.com/environment/millions-of-fossil-fuel-dollars-are-being-pumped-into-anticlimate-lobbying/.

36 Eliot Whittington, 'How Big Are Fossil Fuel Subsidies?' Cambridge Institute for Sustainability Leadership, https://www.cisl.cam.ac.uk/business-action/low-carbon-transformation/eliminating-fossil-fuel-subsidies/how-big-are-fossil-fuel-subsidies.

37 Global Studies Initiative, 'What We Do: Fossil Fuel Subsidies and Climate Change', International Institute for Sustainable Development, https://www.iisd.org/gsi/what-we-do/focus-areas/renewable-energy-subsidies-fossil-fuel-phase-out.

38 Mark Carney, 'Breaking the Tragedy of the Horizon – Climate Change and Financial Stability', speech given at

Lloyd's of London, 29 September 2015, https://www.fsb.org/
wp-content/uploads/Breaking-the-Tragedy-of-the-Horizon-
%E2%80%93-climate-change-and-financial-stability.pdf.

39 The official website for the Network for Greening the
Financial System is https://www.ngfs.net/en. See *A Call
for Action: Climate Change as a Source of Financial
Risk* (NGFS, April 2019), www.banque-france.
fr/en/financial-stability/international-role/
network-greening-financial-system.

40 Moody's, 'Moody's Acquires RiskFirst, Expanding Buy-
Side Analytics Capabilities', press release, 25 July 2019,
https://ir.moodys.com/news-and-financials/press-releases/
press-release-details/2019/Moodys-Acquires-RiskFirst-
Expanding-Buy-Side-Analytics-Capabilities/default.aspx.

41 Fatih Birol, 'Renewables 2018: Market Analysis and Forecast
from 2018 to 2023', International Energy Agency, October
2018, https://www.iea.org/renewables2018/.

42 RE100, 'Companies', http://there100.org/companies.

43 David Roberts, 'Utilities Have a Problem: The
Public Wants 100% Renewable Energy, and
Quick', Vox, 11 October 2018, https://www.vox.
com/energy-and-environment/2018/9/14/17853884/
utilities-renewable-energy-100-percent-public-opinion.

44 Stefan Jungcurt, 'IRENA Report Predicts All Forms of
Renewable Energy Will Be Cost Competitive by 2020', SDG
Knowledge Hub, 16 January 2018, http://sdg.iisd.org/news/
irena-report-predicts-all-forms-of-renewable-energy-will-
be-cost-competitive-by-2020/.

45 United Nations Climate Change, 'IPCC Special Report on Global Warming of 1.5 °C', United Nations Framework Convention on Climate Change, https://unfccc.int/ topics/science/workstreams/cooperation-with-the-ipcc/ ipcc-special-report-on-global-warming-of-15-degc.

46 Sunday Times Driving, '10 Electric Cars with 248 Miles or More Range to Buy Instead of a Diesel or Petrol', *Sunday Times* (UK), 1 July 2019, https://www.driving.co.uk/news/10-electric-cars-248-miles-range-buy-instead-diesel-petrol/.

47 Christine Negroni, 'How Much of the World's Population Has Flown in an Airplane?' *Air & Space*, 6 January 2016, https://www.airspacemag.com/daily-planet/how-much-worlds-population-has-flown-airplane-180957719/; original analysis was carried out by Tom Farrier, an air safety specialist, on Quora: Farrier, 'What Percent of the World's Population Will Fly in an Airplane in Their Lives?' Quora, 13 December 2013, https://www.quora.com/ What-percent-of-the-worlds-population-will-fly-in-an-airplane-in-their-lives.

48 Liz Goldman and Mikaela Weisse, 'Technical Blog: Global Forest Watch's 2018 Data Update Explained', Global Forest Watch, 25 April 2019, https://blog.globalforestwatch.org/ data-and-research/technical-blog-global-forest-watchs-2018-data-update-explained; Gabriel daSilva, 'World Lost 12 Million Hectares of Tropical Forest in 2018', Ecosystem Marketplace, 25 April 2019, https://www.ecosystemmarketplace.com/articles/ world-lost-12-million-hectares-tropical-forest-2018/.

49 Rhett A. Butler, 'Beef Drives 80% of Amazon Deforestation',
    Mongabay, 29 January 2009, https://news.mongabay.
    com/2009/01/beef-drives-80-of-amazon-deforestation/;
    full report here: Greenpeace Amazon, 'Amazon Cattle
    Footprint, Mato Grosso: State of Destruction', February
    2010, https://www.greenpeace.org/usa/wp-content/uploads/
    legacy/Global/usa/report/2010/2/amazon-cattle-footprint.
    pdf.

50 Herton Escobar, 'Deforestation in the Amazon Is Shooting
    Up, but Brazil's President Calls the Data "a Lie"', *Science*,
    28 July 2019, https://www.sciencemag.org/news/2019/07/
    deforestation-amazon-shooting-brazil-s-president-calls-data-
    lie.

51 Yuna He, Xiaoguang Yang, Juan Xia, Liyun Zhao, and
    Yuexin Yang, 'Consumption of Meat and Dairy Products
    in China: A Review', *Proceedings of the Nutrition Society*
    75, no. 3 (August 2016): 385–91, https://doi.org/10.1017/
    S0029665116000641.

52 David Tilman, Michael Clark, David R. Williams, *et al.*,
    'Future Threats to Biodiversity and Pathways to Their
    Prevention', *Nature* 546, (1 June 2017): 73–81, https://www.
    nature.com/articles/nature22900; Jonathan A. Foley, Navin
    Ramankutty, Kate A. Brauman, *et al.*, 'Solutions for a
    Cultivated Planet', *Nature* 478 (2011): 337–42, https://www.
    nature.com/articles/nature10452.

53 EATForum, 'The EAT-Lancet Commission on Food,
    Planet, Health', https://eatforum.org/eat-lancet-commission/.

54 Jean-Francois Bastin, Yelena Finegold, Claude Garcia,

et al., 'The Global Tree Restoration Potential', *Science* 365, no. 6448 (5 July 2019): 76–79, https://science.sciencemag.org/content/365/6448/76.

55 Ibid.

56 World Agroforestry, 'New Look at Satellite Data Quantifies Scale of China's Afforestation Success', press release, 5 May 2017, https://www.worldagroforestry.org/news/new-look-satellite-data-quantifies-scale-chinas-afforestation-success.

57 United Nations Environment Programme, 'Ethiopia Plants over 350 Million Trees in a Day, Setting New World Record', 2 August 2019, https://www.unenvironment.org/news-and-stories/story/ethiopia-plants-over-350-million-trees-day-setting-new-world-record.

58 Roland Ennos, 'Can Trees Really Cool Our Cities Down?' The Conversation, 22 December 2015, http://theconversation.com/can-trees-really-cool-our-cities-down-44099.

59 Amy Fleming, 'The Importance of Urban Forests: Why Money Really Does Grow on Trees', *Guardian* (US edition), 12 October 2016, https://www.theguardian.com/cities/2016/oct/12/importance-urban-forests-money-grow-trees.

60 Humans' meat consumption has varied throughout history but has generally been much lower than at present. Prehistoric humans ate occasional scavenged carrion, while ancient Greeks and Romans consumed between 20 and 30 kilograms per person per year. In the middle ages, European consumption stood at 40 kilograms per capita per year, and in the post-plague

Renaissance, at 110 kilograms. During the Industrial Revolution the average dropped to only 14 kilograms per person per year. See Tomorrow Today, 'A History of Meat Consumption', video, Deutsche Welle, 18 January 2019, https://www.dw.com/en/a-history-of-meat-consumption/av-47130648. Post-industrialisation and refrigeration, meat consumption has steadily increased: from 20 kilograms per person globally in 1960 to 40 kilograms per person globally today. Consumption is highest across high-income countries (with the greatest meat-eaters residing in Australia, consuming around 116 kilograms per person in 2013). The average European and North American consumes nearly 80 kilograms and more than 110 kilograms, respectively. (Hannah Ritchie and Max Roser, 'Meat and Dairy Production', Our World in Data, August 2017, https://ourworldindata.org/meat-and-seafood-production-consumption.)

61 Areeba Hasan, 'Signal of Change: AT Kearney Expects Alternative Meats to Make Up 60% Market in 2040', Futures Centre, 16 July 2019, https://www.thefuturescentre.org/signals-of-change/224145/kearney-expects-alternative-meats-make-60-market-2040.

62 Paul Armstrong, 'Greenpeace, Nestlé in Battle over Kit Kat Viral', CNN, 20 March 2010, http://edition.cnn.com/2010/WORLD/asiapcf/03/19/indonesia.rainforests.orangutan.nestle/index.html.

63 Greenpeace International, 'Nestlé Promise Inadequate to Stop Deforestation for Palm Oil', press release,

14 September 2018, https://www.greenpeace.org/
international/press-release/18400/nestle-promise-
inadequate-to-stop-deforestation-for-palm-oil/. For
further analysis of Nestlé's predicament and its response,
see Aileen Ionescu-Somers and Albrecht Enders, 'How
Nestlé Dealt with a Social Media Campaign against It',
Financial Times, 3 December 2012, https://www.ft.com/
content/90dbff8a-3aea-11e2-b3f0-00144feabdc0.

64 Two extremely useful articles on this subject are
Jonathan Rowe and Judith Silverstein, 'The GDP Myth',
JonathanRowe.org, http://jonathanrowe.org/the-gdp-myth,
originally published in *Washington Monthly,* 1 March 1999;
and Stephen Letts, 'The GDP Myth: The Planet's Measure
for Economic Growth Is Deeply Flawed and Outdated', ABC.
net.au, 2 June 2018, https://www.abc.net.au/news/2018-06-02/
gdp-flawed-and-out-of-date-why-still-use-it/9821402.

65 United Nations, 'About the Sustainable Development
Goals', https://www.un.org/sustainabledevelopment/
sustainable-development-goals/. These goals are: No
Poverty; Zero Hunger; Good Health and Well-being;
Quality Education; Gender Equality; Clean Water
and Sanitation; Affordable and Clean Energy; Decent
Work and Economic Growth; Industry, Innovation,
and Infrastructure; Reducing Inequality; Sustainable
Cities and Communities; Responsible Consumption and
Production; Climate Action; Life Below Water; Life on
Land; Peace, Justice, and Strong Institutions; Partnerships
for the Goals.

66 Dieter Holger, 'Norway's Sovereign-Wealth Fund Boosts Renewable Energy, Divests Fossil Fuels', *Wall Street Journal*, 12 June 2019, https://www.wsj.com/articles/norways-sovereign-wealth-fund-boosts-renewable-energy-divests-fossil-fuels-11560357485.

67 350.org, '350 Campaign Update: Divestment', https://350.org/350-campaign-update-divestment/.

68 Chris Mooney and Steven Mufson, 'How Coal Titan Peabody, the World's Largest, Fell into Bankruptcy', *Washington Post*, 13 April 2016, https://www.washingtonpost.com/news/energy-environment/wp/2016/04/13/coal-titan-peabody-energy-files-for-bankruptcy/.

69 350.org, 'Shell Annual Report Acknowledges Impact of Divestment Campaign', press release, 22 June 2018, https://350.org/press-release/shell-report-impact-of-divestment/.

70 Ceri Parker, 'New Zealand Will Have a New "Well-being Budget," Says Jacinda Ardern', *World Economic Forum*, 23 January 2019, https://www.weforum.org/agenda/2019/01/new-zealand-s-new-well-being-budget-will-fix-broken-politics-says-jacinda-ardern/.

71 Enter Costa Rica, 'Costa Rica Education', https://www.entercostarica.com/travel-guide/about-costa-rica/education.

72 World Bank, 'Accounting Reveals That Costa Rica's Forest Wealth Is Greater Than Expected', 31 May 2016, https://www.worldbank.org/en/news/feature/2016/05/31/accounting-reveals-that-costa-ricas-forest-wealth-is-greater-than-expected.

73 See http://happyplanetindex.org/countries/costa-rica.

74 For a helpful introduction to AI, see Snips, 'A 6-Minute Intro to AI', https://snips.ai/content/intro-to-ai/#ai-metrics.

75 David Silver and Demis Hassabis, 'AlphaGo Zero: Starting from Scratch', DeepMind, 18 October 2017, https://deepmind.com/blog/alphago-zero-learning-scratch/.

76 DeepMind, https://deepmind.com/.

77 Rupert Neate, 'Richest 1% Own Half the World's Wealth, Study Finds', *Guardian* (US edition), 14 November 2017, https://www.theguardian.com/inequality/2017/nov/14/worlds-richest-wealth-credit-suisse.

78 Amy Sterling, 'Millions of Jobs Have Been Lost to Automation. Economists Weigh In on What to Do about It', *Forbes*, 15 June 2019, https://www.forbes.com/sites/amysterling/2019/06/15/automated-future/.

79 Trading Economics, 'Brazil – Employment in Agriculture (% of Total Employment)', https://tradingeconomics.com/brazil/employment-in-agriculture-percent-of-total-employment-wb-data.html.

80 For more information, see Olivia Gagan, 'Here's How AI Fits into the Future of Energy', World Economic Forum, 25 May 2018, https://www.weforum.org/agenda/2018/05/how-ai-can-help-meet-global-energy-demand.

81 David Rolnick, Priya L. Donti, Lynn H. Kaack, *et al.*, 'Tackling Climate Change with Machine Learning', Arxiv, 10 June 2019, https://arxiv.org/pdf/1906.05433.pdf.

82 PricewaterhouseCoopers, 'What Doctor? Why AI and Robotics Will Define New Health', 11 April 2017, https://

www.pwc.com/gx/en/industries/healthcare/publications/
ai-robotics-new-health/ai-robotics-new-health.pdf.

83 Nicolas Miailhe, 'AI & Global Governance: Why We Need an
Intergovernmental Panel for Artificial Intelligence', United
Nations University Centre for Policy Research, 10 December
2018, https://cpr.unu.edu/ai-global-governance-why-we-need-
an-intergovernmental-panel-for-artificial-intelligence.html.

84 Tom Simonite, 'Canada, France Plan Global Panel to Study
the Effects of AI', *Wired,* 6 December 2018, https://www.
wired.com/story/canada-france-plan-global-panel-study-ai/.

85 Richard Evans and Jim Gao, 'DeepMind AI
Reduces Google Data Centre Cooling Bill by 40%',
DeepMind, 20 July 2016, https://deepmind.com/blog/
deepmind-ai-reduces-google-data-centre-cooling-bill-40/.

86 United Nations Division for the Advancement of Women
(UNDAW), 'Equal Participation of Women and Men in
Decision-Making Processes, with Particular Emphasis on
Political Participation and Leadership', report of the Expert
Group Meeting, 24–25 October 2005; Kathy Caprino, 'How
Decision-Making Is Different Between Men and Women
and Why It Matters in Business', *Forbes,* 12 May 2016,
https://www.forbes.com/sites/kathycaprino/2016/05/12/
how-decision-making-is-different-between-men-and-
women-and-why-it-matters-in-business/; Virginia Tech,
'Study Finds Less Corruption in Countries Where More
Women Are in Government', ScienceDaily, 15 June 2018,
https://www.sciencedaily.com/releases/2018/06/
180615094850.htm.

87 United Nations Climate Change News, '5 Reasons
    Why Climate Action Needs Women', United Nations
    Framework Convention on Climate Change, 2 April
    2019, https://unfccc.int/news/5-reasons-why-climate-
    action-needs-women; Emily Dreyfuss, 'Here's a Way
    to Fight Climate Change: Empower Women', *Wired*,
    3 December 2018, https://www.wired.com/story/
    heres-a-way-to-fight-climate-change-empower-women/.

88 Thais Compoint, '10 Key Barriers for Gender Balance
    (Part 2 of 3)', Déclic International, 5 March 2019, https://
    declicinternational.com/key-barriers-gender-balance-2/.

89 Anne Finucane and Anne Hidalgo, 'Climate Change
    Is Everyone's Problem. Women Are Ready to Solve It',
    Fortune, 12 September 2018, https://fortune.com/2018/09/12/
    climate-change-sustainability-women-leaders/.

90 Project Drawdown.

91 Ibid.

92 Brand New Congress, https://brandnewcongress.org/.

93 Andrea González-Ramírez, 'The Green New Deal
    Championed by Alexandria Ocasio-Cortez Gains
    Momentum', Refinery29, 7 February 2019, https://www.
    refinery29.com/en-us/2018/12/219189/alexandria-ocasio-
    cortez-green-new-deal-climate-change; on female solidarity
    and the recognition of US female politicians for the
    suffragist movement: Sirena Bergman, 'State of the Union:
    How Congresswomen Used Their Outfits to Make a
    Statement at Trump's Big Address', *Independent*, 6 February
    2019, https://www.independent.co.uk/life-style/women/

trump-state-union-women-ocasio-cortez-pelosi-suffragette-
white-a8765371.html.

94 Natural Resources Defense Council, 'Salt of the Earth,
Courtesy of the Sun', 30 January 2019, https://www.nrdc.org/
stories/salt-earth-courtesy-sun.

95 Solar Sister, https://solarsister.org.

96 Laurie Goering, 'Climate Pressures Threaten Political
Stability – Security Experts', Reuters, 24 June 2015, https://
uk.reuters.com/article/climatechange-security-politics/
climate-pressures-threaten-political-stability-security-
experts-idUKL8N0ZA2H220150624.

97 Laura McCamy, 'Companies Donate Millions to Political
Causes to Have a Say in the Government – Here Are 10 That
Have Given the Most in 2018', *Business Insider France,* 13
October 2018, http://www.businessinsider.fr/us/companies-are-
influencing-politics-by-donating-millions-to-politicians-2018-9.

98 Influence Map, 'National Association of Manufacturers
(NAM)', https://influencemap.org/influencer/National-
Association-of-Manufacturing-NAM.

99 On the United States, for example, see Andy
Stone, 'Climate Change: A Real Force in the
2020 Campaign?' *Forbes,* 25 July 2019, https://
www.forbes.com/sites/andystone/2019/07/25/
climate-change-a-real-force-in-the-2020-campaign/.

100 For more on Extinction Rebellion, see their website,
https://rebellion.earth/; Brian Doherty, Joost de
Moor, and Graeme Hayes, 'The "New" Climate
Politics of Extinction Rebellion?' openDemocracy, 27

November 2018, https://www.opendemocracy.net/en/
new-climate-politics-of-extinction-rebellion/.

101 For more resources on civil disobedience, see 'Civil
Disobedience', ScienceDirect, https://www.sciencedirect.
com/topics/computer-science/civil-disobedience.

102 Erica Chenoweth, 'The "3.5% Rule": How a Small Minority
Can Change the World', Carr Centre for Human Rights
Policy, 14 May 2019, https://carrcenter.hks.harvard.edu/
news/35-rule-how-small-minority-can-change-world.

103 Fridays for Future, https://www.fridaysforfuture.org/.

104 Jonathan Watts, '"Biggest Compliment Yet": Greta Thunberg
Welcomes Oil Chief's "Greatest Threat" Label', *Guardian*
(US edition), 5 July 2019, https://www.theguardian.com/
environment/2019/jul/05/biggest-compliment-yet-greta-
thunberg-welcomes-oil-chiefs-greatest-threat-label.

Conclusion: A New Story

1 More on Sputnik from NASA: National Aeronautics and
Space Administration, 'Sputnik and the Dawn of the Space
Age', 10 October 2007, https://history.nasa.gov/sputnik/.

2 An analysis of this speech, 50 years on, can be found
here: Marina Koren, 'What John F. Kennedy's Moon
Speech Means 50 Years Later', *The Atlantic*, 15 July 2019,
https://www.theatlantic.com/science/archive/2019/07/
apollo-moon-landing-jfk-speech/593899/.

3 Space Centre Houston, 'Photo Gallery: Apollo-Era
Flight Controllers', 2 July 2019, https://spacecenter.org/
photo-gallery-apollo-era-flight-controllers/.

4   For an analysis of the 'JFK and the janitor' incident and
    what it reveals about inspiration and motivation, see
    Zach Mercurio, 'What Every Leader Should Know about
    Purpose', Huffington Post, 20 February 2017, https://www.
    huffpost.com/entry/what-every-leader-should-know-about-
    purpose_b_58ab103fe4b026a89a7a2e31.

Bibliography and Further Reading

The Problem

Archer, David. *The Long Thaw: How Humans Are Changing the Next 100,000 Years of Earth's Climate.* Princeton, N.J.: Princeton Science Library, 2016.

Carson, Rachel. *Silent Spring.* New York: Mariner Books, 1962.

Masson-Delmotte, V., P. Zhai, H.-O. Pörtner, D. Roberts, J. Skea, P. R. Shukla, A. Pirani, W. Moufouma-Okia, C. Péan, R. Pidcock, S. Connors, J. B. R. Matthews, Y. Chen, X. Zhou, M. I. Gomis, E. Lonnoy, T. Maycock, M. Tignor and T. Waterfield, *eds. Global Warming of 1.5°C. An IPCC Special Report on the Impacts of Global Warming of 1.5°C Above Pre-Industrial Levels and Related Global Greenhouse Gas Emission Pathways, in the Context of Strengthening the Global Response to the Threat of Climate Change, Sustainable Development, and Efforts to Eradicate Poverty.* In press.

Evans, Alex. *The Myth Gap: What Happens When Evidence and Arguments Aren't Enough.* Bodelva, Cornwall, UK: Eden Project Books, 2017.

Ghosh, Amitav. *The Great Derangement: Climate Change and the Unthinkable.* Chicago: University of Chicago Press, 2017.

Goodell, Jeff. *The Water Will Come: Rising Seas, Sinking Cities, and the Remaking of the Civilized World.* New York: Back Bay Books, 2018.

Henson, Robert. *The Rough Guide to Climate Change.* London: Rough Guides, 2011.

Jamail, Dahr. *The End of Ice: Bearing Witness and Finding Meaning in the Path of Climate Disruption.* New York: New Press, 2019.

Jamieson, Dale. *Reason in a Dark Time: Why the Struggle Against Climate Change Failed—And What It Means for Our Future.* Oxford: Oxford University Press, 2014.

Keeling, Charles. 'The Concentration and Isotopic Abundances of Carbon Dioxide in the Atmosphere', *Tellus* 12, no. 2 (1960). https://onlinelibrary.wiley.com/doi/epdf/10.1111/j.2153-3490.1960.tb01300.x.

Hansen, James. *Storms of My Grandchildren: The Truth About the Coming Climate Catastrophe and Our Last Chance to Save Humanity.* New York: Bloomsbury USA, 2010.

Kolbert, Elizabeth. *Field Notes from a Catastrophe: Man, Nature, and Climate Change.* New York: Bloomsbury, 2015.

Lancaster, John. *The Wall: A Novel*. New York: W. W. Norton, 2019.

Lynas, Mark. *Six Degrees: Our Future on a Hotter Planet*. Boone, Iowa: National Geographic, 2008.

Moellendorf, Darrell. 'Progress, Destruction, and the Anthropocene', *Social Philosophy and Policy* 34, no. 2 (2017): 66–88.

Wallace-Wells, David. *The Uninhabitable Earth: Life After Warming*. New York: Tim Duggan Books, 2019.

Designing the Future: Political, Social, Technological and Cultural Change

Davey, Edward. *Given Half a Chance: Ten Ways to Save the World*. London: Unbound, 2019.

Franklin, Daniel. *Mega Tech: Technology in 2050*. London: Economist Books, 2017.

Gold, Russell. *Superpower: One Man's Quest to Transform American Energy*. New York: Simon and Schuster, 2019.

Harvey, Hal. *Designing Climate Solutions: A Policy Guide for Low-Carbon Energy*. Washington, D.C.: Island Press, 2018.

Hawken, Paul, ed. *Drawdown: The Most Comprehensive Plan Ever Proposed to Reverse Global Warming*. London: Penguin Books, 2017.

Latour, Bruno. *Down to Earth: Politics in the New Climate Regime*. Cambridge, UK: Polity Press, 2018.

Leicester, Graham. *Transformative Innovation: A Guide to Practice and Policy*. Charmouth, UK: Triarchy Press, 2016.

Lovelock, James. *The Vanishing Face of Gaia: A Final Warning*. London: Penguin, 2010.

McKibben, Bill. *Falter: Has the Human Game Begun to Play Itself Out?* New York: Henry Holt, 2019.

O'Hara, Maureen, and Graham Leicester. *Dancing at the Edge, Competence, Culture and Organisation in the 21st Century*. Charmouth, UK: Triarchy Press, 2012.

Robinson, Mary. *Climate Justice: Hope, Resilience, and the Fight for a Sustainable Future*. London: Bloomsbury, 2018.

Sachs, Jeffrey D. *The Age of Sustainable Development*. New York: Columbia University Press, 2015.

Sahtouris, Elisabet. *Gaia: The Story of Earth and Us*. Scotts Valley, Calif.: CreateSpace Independent Publishing Platform, 2018.

Smith, Bren. *Eat Like a Fish: My Adventures as a Fisherman Turned Restorative Ocean Farmer*. New York: Knopf, 2019.

Snyder, Timothy. *On Tyranny: Twenty Lessons from the Twentieth Century*. New York: Tim Duggan Books, 2017.

Wahl, Daniel Christian. *Designing Regenerative Cultures*. Charmouth, UK: Triarchy Press, 2016.

Walsh, Bryan. *End Times: A Brief Guide to the End of the World*. London: Hachette Books, 2019.

Wheatley, Margaret J. *Leadership and the New Science: Discovering Order in a Chaotic World*. Oakland, Calif.: Berrett-Koehler, 2006.

Economics

Assadourian, Erik. 'The Rise and Fall of Consumer Cultures', In Worldwatch Institute, ed., *State of the World 2010: Transforming Cultures from Consumerism to Sustainability*. New York: W. W. Norton, 2010.

Jackson, Tim. *Prosperity Without Growth: Economics for a Finite Planet*. London: Routledge Earthscan, 2009.

Klein, Naomi. *On Fire: The (Burning) Case for a Green New Deal*. New York: Simon and Schuster, 2019.

Klein, Naomi. *This Changes Everything: Capitalism vs. the Climate*. New York: Simon and Schuster, 2015.

Lovins, L. Hunter, Stewart Wallis, Anders Wijkman and John Fullerton. *A Finer Future: Creating an Economy in Service to Life*. Philadelphia: New Society, 2018.

Meadows, Donella H., Dennis L. Meadows, Jørgen Randers and William W. Behrens III. *Limits to Growth: The 30-Year Update*. Chelsea, Vt.: Chelsea Green, 2004.

Nordhaus, William. *The Climate Casino: Risk, Uncertainty, and Economics for a Warming World*. New Haven, Conn.: Yale University Press, 2015.

Raworth, Kate. *Doughnut Economics: Seven Ways to Think Like a 21st-Century Economist*. New York: Random House, 2017.

Rowland, Deborah. *Still Moving: How to Lead Mindful Change*. New York: Wiley Blackwell, 2017.

Personal Action and Movement Building

Bateson, Gregory. *Steps to an Ecology of Mind*. New York: Chandler, 1972.

Berners-Lee, Mike. *There Is No Planet B: A Handbook for the Make or Break Years*. Cambridge: Cambridge University Press, 2019.

Extinction Rebellion. *This Is Not a Drill: An Extinction Rebellion Handbook*. London: Penguin, 2019.

Foer, Jonathan Safar. *We Are the Weather: Saving the Planet Begins at Breakfast*. New York: Farrar, Straus and Giroux, 2019.

Friedman, Thomas L. *Thank You for Being Late: An Optimist's Guide to Thriving in the Age of Acceleration*. New York: Farrar, Straus and Giroux, 2016.

Havel, Václav. *Disturbing the Peace: A Conversation with Karel Huizdala*. New York: Vintage Books, 1991.

Louv, Richard. *Last Child in the Woods: Saving Our Children from Nature-Deficit Disorder*. New York: Algonquin, 2005.

Macy, Joanna, and Chris Johnstone. *Active Hope: How to Face the Mess We're in Without Going Crazy*. San Francisco: New World Library, 2012.

Mandela, Nelson. *A Long Walk to Freedom*. New York: Time Warner Books, 1995.

Martinez, Xiuhtezcatl. *We Rise: The Earth Guardians Guide to Building a Movement that Restores the Planet*. New York: Rodale Books, 2018.

Quinn, Robert E. *Building the Bridge As You Walk on It: A Guide for Leading Change*. Greensboro, N.C.: Jossey-Bass, 2004.

Plous, Scott. *The Psychology of Judgment and Decision Making*. Philadelphia: Temple University Press, 1993.

Scranton, Roy. *Learning to Die in the Anthropocene: Reflections on the End of Civilization*. San Francisco: City Lights, 2015.

Sharpe, Bill. *Three Horizons: The Patterning of Hope*. Charmouth, UK: Triarchy Press, 2013.

Seligman, Martin E. P. *Learned Optimism: How to Change Your Mind and Your Life*. London: Vintage, 2006.

Solnit, Rebecca. *Hope in the Dark: Untold Histories, Wild Possibilities*. Chicago: Haymarket Books, 2016.

Thunberg, Greta. *No One Is Too Small to Make a Difference*. London: Penguin, 2019.

Wheatley, Margaret J. *Who Do We Choose to Be? Facing Reality, Claiming Leadership, Restoring Sanity*. Oakland, Calif.: Berrett-Koehler, 2017.

Nature

Baker, Nick. *ReWild: The Art of Returning to Nature*. London: Aurum, 2017.

Brown, Gabe. *Dirt to Soil: One Family's Journey into Regenerative Agriculture*. London: Chelsea Green, 2018.

Eisenstein, Charles. *Climate: A New Story*. Berkeley: North Atlantic Books, 2018.

Glassley, William E. *A Wilder Time: Notes from a Geologist at the Edge of the Greenland Ice*. New York: Bellevue Literary Press, 2018.

Kolbert, Elizabeth. *The Sixth Extinction: An Unnatural History*. London: Picador, 2015.

Monbiot, George. *Feral: Rewilding the Land, Sea and Human Life*. London: Penguin, 2015.

Oakes, Lauren E. *In Search of the Canary Tree: The Story of a Scientist, a Cypress, and a Changing World*. New York: Basic Books, 2018.

Simard, Suzanne. *Finding the Mother Tree*. London: Penguin Random House, 2020.

Tree, Isabella. *Wilding: The Return of Nature to a British Farm*. London: Picador, 2018.

Wohlleben, Peter. *The Hidden Life of Trees: What They Feel, How They Communicate – Discoveries from a Secret World*. Vancouver, B.C.: Greystone Books, 2016.

Wulf, Andrea. *The Invention of Nature: Alexander von Humboldt's New World*. New York: Vintage, 2015.